T5-BZZ-807

PERGAMON INTERNATIONAL LIBRARY
of Science, Technology, Engineering and Social Studies
The 1000-volume original paperback library in aid of education, industrial training and the enjoyment of leisure
Publisher: Robert Maxwell, M.C.

SOLAR VERSUS NUCLEAR

Choosing Energy Futures

Other Titles of Interest

GABOR *et al*	Beyond the Age of Waste
GVISHIANI *et al*	Science, Technology and Global Problems
HOWELL	Your Solar Energy Home
HUNT	Fission, Fusion and the Energy Crisis, 2nd Edition
MAROIS	Towards a Plan of Actions for Mankind (5 vols.)
McVEIGH	Sun Power
MONTBRIAL	Energy: The Countdown
MURRAY	Nuclear Energy, 2nd Edition
SECRETARIAT FOR FUTURES STUDIES	Resources, Society and the Future
SIMEONS	Coal: Its Role in Tomorrow's Technology
SIMON	Energy Resources
STARR	Current Issues in Energy
UNECE	Coal: 1985 and Beyond
UNITAR	Future Supply of Nature-Made Petroleum and Gas
VELIKHOV *et al*	Science, Technology and the Future
WILLIAMS	Carbon Dioxide, Climate and Society

SOLAR VERSUS NUCLEAR
Choosing Energy Futures

A Report Prepared for the Swedish Secretariat for Futures Studies by

MANS LÖNNROTH
THOMAS B. JOHANSSON
PETER STEEN

Translated from the Swedish by
P. C. HOGG

PERGAMON PRESS
OXFORD · NEW YORK · TORONTO · SYDNEY · PARIS · FRANKFURT

U.K.	Pergamon Press Ltd., Headington Hill Hall, Oxford OX3 0BW, England
U.S.A.	Pergamon Press Inc., Maxwell House, Fairview Park, Elmsford, New York 10523, U.S.A.
CANADA	Pergamon of Canada, Suite 104, 150 Consumers Road, Willowdale, Ontario M2J 1P9, Canada
AUSTRALIA	Pergamon Press (Aust.) Pty. Ltd., P.O. Box 544, Potts Point, N.S.W. 2011, Australia
FRANCE	Pergamon Press SARL, 24 rue des Ecoles, 75240 Paris, Cedex 05, France
FEDERAL REPUBLIC OF GERMANY	Pergamon Press GmbH, 6242 Kronberg-Taunus, Pfderdstrasse 1, Federal Republic of Germany

First edition 1980

British Library Cataloguing in Publication Data

Solar Versus Nuclear: Choosing Energy Futures.
(Pergamon International Library)
I. Atomic power - Sweden
2. Solar energy - Sweden
I. Title II. Johansson, Thomas B
III. Steen, Peter IV. Secretariat for
Futures Studies
333.7 HD9502.S/ 79-40725

ISBN 0-08-024758-X Hardcover
ISBN 0-08-024759-8 Flexicover

In order to make this volume available as economically and as rapidly as possible the authors' typescripts have been reproduced in their original forms. This method unfortunately has its typographical limitations but it is hoped that they in no way distract the reader.

Printed and bound in Great Britain by
William Clowes (Beccles) Limited, Beccles and London

Contents

Preface

By decision of June 30, 1971, the Prime Minister appointed a Working Party under the chairmanship of Mrs. Alva Myrdal, Cabinet Minister, to deal with questions of future studies in Sweden.

The report of the Working Party, "To Choose a Future - a basis for discussion and deliberations on futures studies in Sweden", was submitted to the Prime Minister on August 25, 1972.

Following the proposal of the Working group a secretariat for futures studies was set up in 1973 by the Swedish government. The present book - Sweden beyond oil - nuclear commitments and solar options - is the final report of the second completed futures studies project. The perspective for these studies is more accurately explained on page 166.

Introduction

If there was ever a country for which nuclear energy seemed ideal, Sweden was the one. It has high standards of living, energy intensive industries, advanced technology in many fields, strong tradition of electrification and a competent utility organization, abundant reserves of (admittedly low grade) uranium, low population density and a large number of suitable reactor sites. Moreover Sweden has no fossil fuels at all. All those factors made nuclear power the obvious choice for the future during the 50's, when the first crucial decisions were taken. In the beginning of the 70's nuclear energy seemed well established with a Swedish reactor industry, in fact the only light water reactor-design outside the communist countries developed without U.S. licenses. The utilities were firmly committed and there were advanced plans of heating the major metropolitan areas in Stockholm, Gothenburg and Malmo by district heating supplied from nuclear heat and power plants. The Swedish plans for nuclear generation of electricity were the most ambitious in the world, on a basis of installed capacity per capita. All decisions had been taken with large majorities in the Parliament, and gradually a more powerful nuclear establishment was built up.

Then the roof fell in. A heated debate erupted in 1974 and can be traced back to 1972. By the end of 1978 two governments had been brought down on the issue of nuclear power. This is not the place for explaining the process of politicization of nuclear power, only to state what now is a fact. Nuclear power is viewed by large parts of the public as something much more than just "another way of boiling water". Concern for the economy or the environment, although important, cannot fully explain the degree of controversy. In my mind it is obvious that nuclear energy - deservedly or not - has become a symbol for something more profound than economic growth or environmental decay, the social impact on society or super-high and rapidly changing technology.

The nuclear controversy can thus be seen as a controversy as much over the distribution of power over technological choice in general as over a particular technology. Elites of all nations do not readily accept challenges to their power. All the more noteworthy, therefore, is the initiative of the Swedish government in 1973 to establish a secretariat for futures studies.

This secretariat was given the explicit role to study independently the long-term implications of present decisions as well as alternative development options. Needless to say such an initiative has not been uncontroversial - there are many inside the government machinery who have seen the work of the secretariat as a thorn-in-the-flesh, an undeserved nuisance.

This book "Sweden beyond oil - nuclear commitments and solar options" brings out some of the more important challenges of the nuclear contoversy, namely an analysis of the mechanisms of technological choice and the role of industry, agencies and other groups of special interests and roles vis-a-vis, the democratically elected and appointed bodies in local and central governments.

It is my personal view that the width of this work presented here by contrast demonstrates rather vividly the narrowness of most academic disciplines. Knowledge of technical characteristics wedded to organizational sociology, some economics and a dose of political science goes a long way in creating a framework for understanding the politics of technical change that I myself think is needed in many other areas.

Four aspects are worth mentioning in this work. Firstly the need for an historical perspective. It takes decades to develop new technology, and the institutions that today shape the technologies of tomorrow have themselves evolved during several decades. In the Swedish case the institutional framework for electricity can only be understood by looking at Sweden from the beginning of this century.

Secondly the need for detached attitudes to forecasts. It is quite obvious from past experiences that the time horizon over which it is meaningful to make forecasts on the demand for energy is conspiciously shorter that the lead times necessary for introducing and developing new technologies. This is a dilemma which cannot be solved through more ingenious forecasting methods - in my view the problem of projecting GNP over say three to four decades is not only unsolved but also unsolvable. The meaning of say a 3% growth rate from the level of GNP existing e.g. in the U.S. or Sweden is open to such devastating objections that other attempts seem necessary.

The approach used in this work does offer some points of departure. Instead of making elaborate forecasts a minimum choice of assumptions are made. These comprise possible rates of change in the productivity of energy and labour, size of the total labour force etc. together with the need for labour in building up different new energy systems. Thus not a forecast but merely a check of consistency is made, saying in effect nothing more than whether and how the energy systems studies are compatible with a further increase in economic well-being.

What will actually happen is another matter. The aim has been to decide whether solar or nuclear energy can replace imported oil, given the knowledge we have today. The answer is for the two alternatives a, very tentative, yes. There are, however, important areas of uncertainty regarding the long term potential and impacts of the two paths studied here. Needless to say there are other combinations than these. Technological development may bring forward new options and more experience may show some others to have less potential than previously thought. This, however, just underlines the central point in the argument - what is needed is not long term commitments but flexibility and the capacity to adaption and adjustment.

Thirdly this work brings, in my mind rightly, the explicit recognition that powerful vested interests always exist around any one already established technology. The main short-term political advantage of nuclear energy over solar energy is just that. While the latter is nothing but a R&D scheme and a number of loosely coordinated projects in a frequently fragmented (but innovative) industry the former is represented by a coordinated, almost army-like, set of institutions centred around the electric utilities, the electric supply industry , the nuclear reactor industry, and the construction industry. Moreover the nuclear development is stabilized and facilitated by a whole set of rules, regulations and laws, some explicitly designed for this purpose but most of them just being there for other reasons. The advantage of nuclear energy has to no small degree been just that its entrepreneurs have been skillful enough both to use the existing legal and institutional framework and at times powerful enough to change and adjust it when the obstacles have been too great.

As the more detailed analysis shows, the framework that is beneficial to nuclear energy frequently puts solar energy (and to some extent conservation) at a disadvantage. Hence the fourth aspect to be stressed is, as I see it, the analysis of how the institutional framework has to be changed in order to open up the potential for the solar industries. As is shown, this institutional rearrangement is no small endeavour- many areas have to be dealt with, and strong interests can and will be marshalled against it. To establish a solar industry - which is necessary if we are to learn the true potential of this supply alternative - therefore requires that a market is opened up. Entrepreneurial interests have to be stimulated by the government and given enough stability so that the inherent risks of technological development can be reduced to manageable proportions. The fact that those measures at the same time tend to destabilize the market conditions for the nuclear industry makes the need for public involvement and governmental planning all the more obvious.

I am convinced that the line of argument here can be generalized to other techno-industrial complexes. The military industry, the automobile industry, telecommunications industry, perhaps pharmaceutical industry, all thrive in a setting in which the national government is at the same time guardian and prisoner of the techno-complex.

The analysis in this book - which obviously can be carried further and made more precise - illustrates the need for thinking about institutions rather than abstract economic models.

While this exercise has a Swedish focus there is no doubt in my mind that many of the main conclusions are valid also in other countries and in an international context.

Perhaps the most interesting question is whether the economic conclusions hold also for other countries. When the cost estimations of nuclear and solar energy that existed at the end of 1977 are used the two alternatives are found to be economically roughly equivalent. When the admittedly large uncertainties for the future surrounding both advanced nuclear technologies (particularly the plutonium cycle) and solar technologies are taken into account a more reasonable conclusion seems to be that it is yet too early to say which alternative is less costly than the other. One needs time to find out, hence the need for a flexible and adjustable energy policy.

This conclusion is, however, itself surprising. The conventional wisdom has always had it that nuclear energy is much less costly than solar energy. And here it is shown that even in a country on the same latitude as Alaska the conventional wisdom may very well be wrong. In the Swedish case the lower solar insolation due to the higher latitude is more than offset by the large areas suitable for biomass production.

Other countries, particularly in Europe, have a somewhat higher solar insolation but do not have the advantage of low population densities. Underdeveloped countries, regularly situated in the tropical or subtropical zones, have substantially higher levels of solar insolation than Sweden, but many of them are very crowded.

It seems to me, therefore, that similar types of studies should be made for other countries. This should give us more important information for the discussion of whether the spread of nuclear energy - and in particular the plutonium economy should and could be halted or not.

There are risks connected with nuclear technology and plutonium dispersed among a large number of countries. Hence, one should think carefully before the world embarks on a major commitment on the plutonium economy. A strategy for buying time and deferring major commercial applications of reprocessing and breeder reactors should be worked out, taking the different situations of different countries into account.

One could think of an international agreement between uranium suppliers and importers, assuring the latter of adequate supply and in return for this requesting the same countries to defer commercialization of plutonium fuel cycles. Sweden could play a useful role both in development of alternatives to nuclear energy and possibly also in uranium supply.

The strongest objection to a strategy for deferring the plutonium economy comes, as usual, from the commercial interests of those that have developed components of the plutonium cycle, partly for domestic reasons, partly as an export potential. It is high time that the national and the international communities realize that the commercial interests, however vital they may seem to those immediately concerned, must not, after all, be mistaken for public interest. There is also high time to think creatively about employment. It is totally unacceptable that the governments of the industrialized countries - and this includes Sweden - should go on sacrificing both long and medium term interest for the short term interest of keeping up employment in this or that armament industry, reprocessing or reactor industry.

Stockholm in January 1979

Gunnar Myrdal

Summary of Contents

All activities presuppose energy. In chapter 1 we deal with some different ways of describing and viewing energy.

Our society has changed its energy sources in the past. Such changes have taken a long time and have affected social functions far beyond the energy sector. In chapter 2 the earlier changes of energy sources are analysed and we attempt to describe the relations between the development of society and energy consumption.

The world and Sweden are today dependent on a few energy sources, above all oil, gas and coal. Some countries get part of their energy supplies from nuclear power and uranium. The durability and other essential characteristics of these energy sources are briefly dealt with in chapter 3.

In chapter 4 two distinct developments are described. One leads to a Nuclear Sweden with its energy supply based chiefly on uranium and the other to a Solar Sweden based on renewable energy sources. We compare the characteristics and costs of the two alternatives and discuss the possibilities of combining them.

In chapter 5 a discussion follows on the possible developmental tendencies built into the present energy policy. There is strong evidence that a number of built-in forces in society are promoting a development towards the nuclear alternative.

In chapter 6 we submit proposals for a conceivable transitional solution for the 1980's which could reduce the dependence on oil while preserving both the solar and nuclear option as possible alternatives for the future.

Forthcoming changes in our energy system will require an element of strong central planning while at the same time the renewable alternatives, in particular will require a locally based organization. Chapter 7 discusses organizational changes in energy conservation and energy production.

In the concluding chapter 8 we take up some questions regarding the effect of the energy systems on the long-term development of society.

Units of measurement

In connection with very large and very small quantities symbols have been introduced. Instead of writing, e.g. a thousand million the symbol (prefix) G (giga) is used.

1GWh = 10^9 Wh = 1,000,000,000 Wh

The following prefixes occur:

Prefix	Numerical factor
E (exa)	10^{18}
P (peta)	10^{15}
T (tera)	10^{12}
G (giga)	10^{9}
M (meta)	10^{6}
k (kilo)	10^{3}
m (milli)	10^{-3}
u (micro)	10^{-6}
n (nano)	10^{-9}
p (pico)	10^{-12}

Example

1 TJ = 10^{12} J = 1,000 GJ

1 Mton = 10^6 tons = 1,000,000 tons

Conversion factors

Energy

1 PJ (petajoule) = 0.2778 TWh (terawatt hours) = 0.0239 Mtoe (million tons of oil equivalents)

1 TWh = 3.6 PJ = 0.0860 Mtoe

1 Mtoe = 41.87 PJ = 11.63 TWh

Diagram for conversion between some different energy units.

1 J = 2.778 x 10^{-7} kWh

1 kWh = 3.6 x 10^6 J

1 GJ = 277.8 kWh = 0.0239 toe

1 toe = 41.87 GJ = 11,630 kWh

Units used for oil, natural gas, wood

1 Mton = 1 million tons

1 Mtoe = energy quantity corresponding to 1 million tons of oil

1 ton of oil = 11.63 MWh energy content (heat)

1 ton of oil = 7.33 barrels of oil (average value, crude oil is of varying density)

1 barrel = 159 litres = 0.159 m^3

1 m^3 of oil = 0.86 ton of oil (average value)

10 $ per barrel = c. 270 Sw Cr/m^3

1 billion m^3 of natural gas = 10^9 m^3 = 0.8 Mtoe (energy content, approximately)

1 ton of wood (dry weight) = c. 0.5 toe = c. 5 MWh

Units used for renewable energy sources

Energy intensity (power per unit of area)

1 W/m^2 = 3.6 kJ/m^2/h = 0.086 cal/cm^2/h (h = hour)

Area

1 ha = 1 hectare = 10,000 m^2

Time

1 year = 8,760 hours

1 Energy - Some Viewpoints and General Approaches

Energy - on the market and in nature

"Let there be light: and there was light."

That is the short and pithy description in the Bible of how nature's great fusion reactors were charged and activated. It may be a timely reminder in the introduction to a book about future energy alternatives. In industrial society man has gradually evolved an energy structure parallel to nature's own. It is debatable whether it has become too large or is still too small, and we can compare it with nature's own energy flow.

In Sweden we used about 30 million tons of oil in 1975. Is that a lot or a little? It is as much oil as was used in India, but it is no more than 1.1% of the 2.7 billion tons which were used throughout the world. Other large energy sources are coal and gas. Every year fossil fuels (oil, coal and gas) are consumed in a total amount equivalent to 5.7 billion tons of oil. Hydroelectric power and nuclear power corresponded to 200 million tons of oil. As a whole, therefore, a quantity of energy is sold and bought in the world every year corresponding to 6 billion tons of oil.

Is that, then, a lot or little? This again depends on what we compare it with. It is almost twice as much as was sold in 1960 and that is quite a lot compared with what is thought to be available - at least this applies to oil and gas. But if we compare it with the amount of energy which circulates in nature we get a different picture. Each year photosynthesis in plants binds an amount of energy which is 8 times greater than the energy that is sold. Winds, waves and currents are driven by an amount of energy which is 60 times greater. The flow of water through the atmosphere - the hydrological circulation - in the form of evaporation and precipitation annually absorbs quantities of energy which are 7,000 times as large as those of the energy market.

But the energy required by the hydrological circulation is, in its turn, only a quarter of the amount which radiates onto the globe from the sun. The major part vanishes into space again in the form of heat radiation or direct reflection. The amount of energy reaching earth from the sun is thus about thirty thousand times as large as that which man puts to use through the various energy sources oil, coal, gas and uranium.

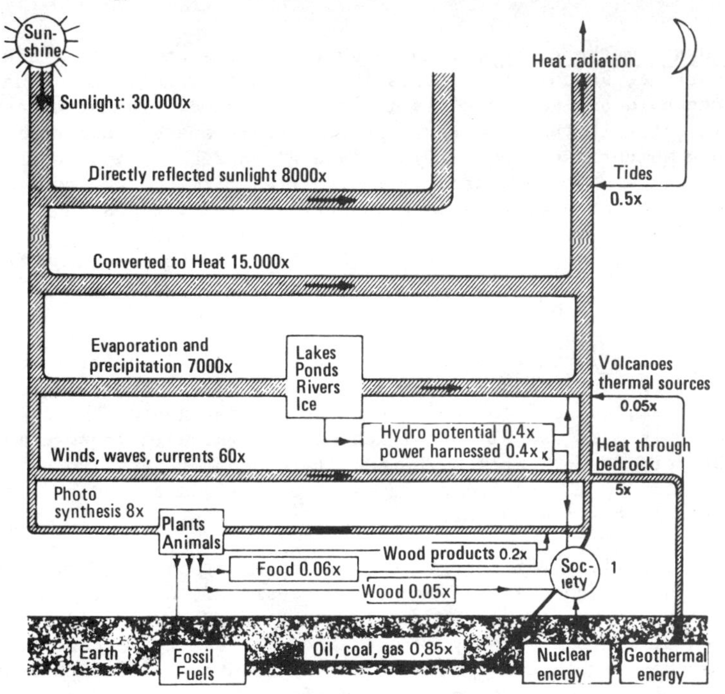

Figure 1. Energy flow on earth. The energy flows are calculated in relation to the energy circulation in society, which has been set at one. Oil, coal and gas, for example, contribute 85% of the latter. That part of the solar energy radiation which is required to drive winds, waves and currents is 60 times as large, etc. (1).

The fusion of atoms in the sun releases the energy which later appears on earth as heat or stored in green plants or fossils. The energy passes through different systems on earth and returns to space as heat radiation. Only a very small part of the energy flow is stored or utilized by man.

From a scientific point of view energy cannot be consumed. It is a flow. By means of various technical aids man can divert it from its flows (or stores), distribute it and make use of it.

We therefore utilize both the natural energy flow and the energy flow created by man in the industrialized system. The former predominates totally in size. If we have the cunning to exploit the natural energy flow for what we regard as necessary applications of energy, the physical amount of energy will not constitute a problem.

From an economic point of view, however, energy is a commodity which is sold by the energy enterprises and therefore appears in statistics as energy consumption.

Here are three distinctions which are worth remembering.

Stored and free-flowing energy sources. The stored energy sources, such as oil, coal and gas, are created from the free-flowing energy sources but at such a slow rate that they can be regarded as non-renewable. The free-flowing energy sources - wind, water, waves, biomass etc - are part of nature's energy flow.

Marketed and non-marketed energy. Only certain forms of energy are sold today. This is important, because the statistics of energy consumption usually record what has been sold and not what has in fact been used. In the case of the developing countries, for instance, a large part of their energy requirement it met with wood and dung which is collected by individual households and never recorded in the energy balance. In industrialized countries like Sweden, to take an example, we utilized the solar radiation through south-facing windows in our houses, while at the same time making use of the energy provided by radiators. But we only count the latter. Another aspect of this is that the sun provides heat from absolute zero-point, i.e. -273° C, to about 0° C, after which domestic heating raises the temperature to about +20° C.

Another distinction is between quantity and quality. Quantities of energy are described in different measures such as kilowatt hours (kWh), megawatt hours (MWh), gigajoules (GJ) and so on(1). In so doing one ignores the fact that 1 kWh of electricity is more useful than 1 kWh of oil, which is in turn more useful than 1 kWh in the form of hot water. The quality concept relates to utility and an essential reason why the stored energy sources (oil, coal and gas) are so widely used is precisely their high quality.

The quality (or utility) of energy is thus dependent on the form in which it appears. In physics another measure is therefore often used, namely exergy (2, 3). Exergy may be said to be a measure of the degree of organization among the particles of matter: exergy is consumed when energy is converted from one form into another. When energy is used it is transferred from a more organized form (e.g. chemical compounds in the form of oil) to a more disorganized one (e.g. hot water at 60° C). Conversely, when one wants to construct better organized forms (e.g. steel) from more disorganized ones (e.g. iron ore) the exergy is incorporated in the new materials.

The exergy concept provides partly new aspects on the use of energy. Thus it links together energy and material resources and illustrates the fact that a society's use of materials is of great significance for its energy needs.

The exergy concept takes account of the fact that not only the amount of energy (the number of MWh) but also the form of energy (electricity, oil, hot water) is significant. In addition the concept makes possible a deeper analysis of efficiency. It thus becomes a natural point of departure for conservation-oriented policies, technological development and organizational change. What is thermodynamically (2) prudent will also in the long run be economically prudent.

Even if a scientific perspective and concepts like exergy and thermodynamics are not particularly easy to translate into concrete decisions, they may yet be regarded as dimensions of energy policy along with - or rather superordinate to - other more traditional concepts such as socio-economic efficiency and profitability Despite this, energy policy must be concretely based on an analysis of how present-day society uses energy.

Energy in society

Can one describe the role of energy in society without at the same time describing society? No, not if the description is to be of significance for the political process relating to the form of energy it supplies. In that respect it is the all-round view that matters - the demand for participation must be matched by

(1) See p. 2, Units of measure
(2) Thermodynamics - the study of energy conversion, of which thermology forms a part.

opportunities for understanding and acquiring insight. But this insight and understanding cannot be automatically derived from the flood of statistics provided by official reports, government bodies and large energy enterprises.

Every discussion of future society generally leads to the formulation of very summary descriptions. This is true both of social development in general and of energy consumption. It can apply to the productive growth of industry or to the total number of terawatt hours (TWh) utilized for transport activities.

The descriptions are a normal tool for the planners of energy supplies. They thereby also automatically become the basis for general political analyses and are a prerequisite for taking decisions on the development of energy production - e.g. how many rivers, nuclear power stations, refineries and coaling ports are to be developed, where land is to be reserved, and so on. At the same time they provide an inadequate "finger-tip feeling" for the effects which the decisions will have on citizens generally. They are therefore not particularly suitable as a basis for the broad political debate. People do not in the first instance ask how many kilowatt hours will be consumed.

The fundamental questions concern actual living conditions: security, work, material standards, the availability of various services, and so on. We therefore also need descriptions which view society and the energy questions more from the level of the individual or of households.

From this level the world usually looks different. One visualises other possibilities for change than when looking at it from the angle of the producer. This is what an American energy researcher (4) writes about the efforts that have been made to see things from the other side. He distinguishes between the viewpoints "upstream" (from the producer) and "downstream" (from the consumer):

"Conservation, from upstream, looks like deprivation. The supplies of energy, believing demand to be insatiable, expect upheaval and dislocation to attend any failure of supply to grow at historical rates. In a perfectly natural way, they see the most essential functions energy permits (keeping warm, having light at night, going quickly wherever one wants) as the functions at peril if they do not do their jobs well.

From downstream, conservation looks quite different. The marginal uses of energy, not the most essential ones, are at stake in the economics of conservation. Conservation is a set of opportunities to be more clever with end-use technology and to attend to portions of the economy where, over many decades, waste has become embedded in the social fabric".

From a general point of view we believe that it is very important to try to describe social development and energy from the "downstream" perspective. But there are not yet suitable data and statistics for such attempts. The studies of living standards that have been conducted in Sweden (5), for example, still only provide "snapshots" and are, naturally enough, not formulated in such a way as to provide information relevant to energy policy.

Figure 2 shows some different ways of describing how Swedish society uses energy. The first describes energy sources and is, as we have said, necessary for the planning of the energy supply. The second describes a distribution of energy consumption between industry, communications, housing and services. Forecasts of energy consumption are sually made in these terms, since that is how society appears from the standpoint of the energy producers. The above description is simply a compilation of the accounts of the energy enterprises. The third way describes energy consumption from a physical point of view, in the form of

Energy saving can be seen from the higher view of the producer as privation but from below as a number of opportunities to act more wisely with the technology at hand. (Photo: Pressens Bild)

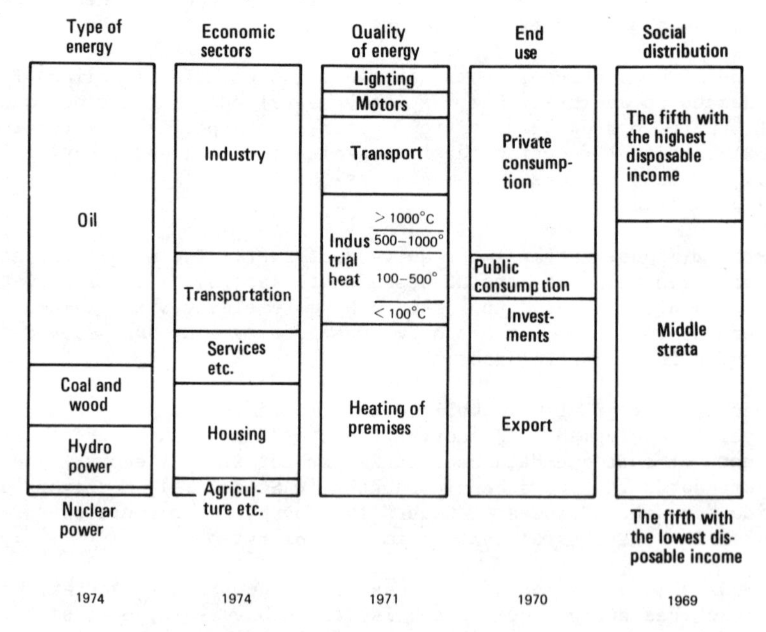

Figure 2. Sweden's energy consumption with breakdown by uses. The dates refer to the year covered by each analysis or study (6, 7, 8, 9).

different qualities of energy. The fourth way attempts to turn the whole pattern around and to see instead, from the perspective of the national economy, how we use energy for private and public consumption, investments and export . The fifth diagram shows roughly how various households use energy. Not surprisingly it turns out that those with higher incomes also consume more energy.

These diagrams provide five angles of approach, but they do not show the connections between them. We shall try to do so, taking as our starting point the second diagram which shows the distribution of energy consumption between different economic sectors.

First we look at industry, where we see large differences between various branches (Table 1).

The differences in energy consumption per worker are very great. Labour-intensive industry consumes relatively little energy. Conversely, those which consume a lot of energy use a small amount of labour power.

This does not mean that energy is easily interchangeable with manpower in the individual production process or that such an interchange has taken place historically. The differences in energy consumption derive from fundamental differences between the industrial processes. In processing industry energy is primarily used for chemical conversion which requires high temperatures, while the engineering industry for example requires power primarily for machining, for motors in machinery and for heating premises.

Table 1. Industrial energy consumption 1971 (10)

Industrial branch	Direct energy TWh	Percentage of industrial energy consumption	Percentage of workers in the industry	Energy consumption per worker MWh
Paper-pulp	62	41	7	1,300
Iron and metal	28	19	6	670
Non-metalliferous mining	13	9	4.5	430
of which cement and lime	7	5	0.3	3,800
Engineering	13	9	41	47
Chemical	11	8	6.4	260
Industry as a whole	150	100	100	233

There are of course connections between different branches. The production of a car (engineering industry) requires sheetmetal (iron and metal industry), plastics (chemical industry) etc. Apart from the direct energy consumption for a particular manufacturing process we must also take into account the energy which goes into producing raw materials and semi-manufactures, etc., which are used in the respective industries. This involves an indirect energy consumption. In the automobile industry, for instance, 37 MWh are required in direct energy consumption per year and employee in order to keep the machines working, heat the premises etc., but in addition 94 MWh are required in direct energy consumption (11) for different materials and so forth.

The shipbuilding industry is usually regarded as a heavy industry and in the public mind heavy industry is associated with high energy consumption. In actual fact the shipbuilding industry - reckoned by direct energy consumption per employee - is about as energy-intensive as, for example, the health service sector. On the other hand the shipbuilding industry does use considerably more indirect energy - for ship plate and other materials. The indirect energy consumption on shipbuilding is more than twice as large as the direct kind.

Analyses also exist of how much energy is consumed at different stages of production (12), e.g. of housing and domestic capital goods. It becomes apparent that it is primarily the production of materials that requires energy. To form and assemble components is relatively economical in energy. For that reason the weight and the materials readily provide some idea of the energy consumption needed to manufacture a product.

From the consumer's point of view many industrial products consume energy when they are put to use. If, for example, we compare how much energy a car, a house or a refrigerator consumes while in use during its lifetime with what it consumes during manufacture, it becomes apparent that the former is usually larger than the latter. A car for instance, consumes about as much energy in the form of petrol in one year as that required for its manufacture, including all materials, transport and so on. To heat a house for a year or two requires more energy than it took to build it. The examples can be multiplied and indicate that it is primarily in the use of commodities that energy is required. This does not mean that the producer has no responsibility - the design of the product largely determines how much energy is required at the stage of use.

One can make similar calculations for all other commodities purchased by households If we add everything together we can calculate how much energy households as a whole consume - whether it is to drive the cars and heat the food or to manufacture cars and food. We can do the same for the public sector (health services, schools, defence etc), for investments and exports, and thus obtain the values shown in table 2.

Table 2 demonstrates that the energy consumption (per krona) varies between different areas of use. Private consumption includes a large amount of commodities which have been imported. The manufacture of these commodities consumes energy abroad, an estimated 100 TWh/year which is not, however, included in the table. Overall it can be stated that private consumption accounts for two thirds to three quarters of the energy used in Sweden (the proportion varies depending on how it is defined).

The separation of private and public consumption is of course rather artificial. It has nothing to do with energy policy or actual living standards. What we generally call private consumption presupposes the activities which we in Sweden have decided to finance in common. This includes roadbuilding, education, health services etc. In many other countries other divisions have been adopted and thus their "private consumption" includes part of that which is "public" with us. Figure 3 is an attempt to show the energy consumption in our particular way of life irrespective of the method of financing. We have calculated it in relative rather than actual figures, as the statistics are 8-10 years old.

The energy consumption of different households has also been studies (9, 14). It emerges that the consumption of energy, direct and indirect, is largely proportional to the disposable income, i.e. salary plus allowances minus tax. The higher the disposable income the higher the consumption of energy.

If we distribute the energy consumption of households between different income groups we obtain the following picture (9).

Table 2. Energy consumption in different sectors 1970 (13). The figures include both direct and indirect energy consumption and labour power.

	Energy consumption		Labour million working hours		kWh/krona
	TWh	%		%	
Private consumption	191	44	2,190	36	3.1
State consumption	19	4	480	8	1.7
Local government	23	5	930	15	1.2
Investments	58	13	1,250	20	2.2
Exports	135	31	1,180	19	4,7
Changes of stocks	14	3	100	2	
Total use	434		6,130		

The 20% of households which have the lowest disposable incomes use 7% of the energy. The 20% with the highest disposable incomes use 37%. The 60% with medium incomes use 56% of the energy.

Those who live in their own homes consume slightly more energy than those who live in blocks of flats. The differences apply both to heating and transport and occur at all income levels. The differences in heating depend mainly on the larger dwelling space. Longer travelling distances and a higher proportion of motorists explain the differences in transport (14). There are also differences between households in large and small urban areas. Households in completely rural districts, for example, use almost 50% more fuel than urban households with equivalent incomes (14).

In addition it appears that education, for instance, at a given income, is of great significance. It seems as if complex characteristics such as activity and

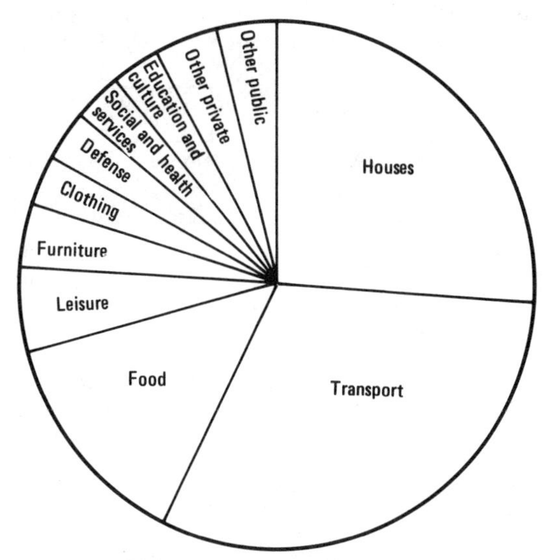

Figure 3. Energy consumption per average household (9).

participation also cause higher energy consumption - with the present-day structure of society.

This naturally applies also to the entry of women into the labour market - with the present structure of the transport system it would be surprising if it did not lead to more two-car households.

Even if this information provides a somewhat better grasp of what energy really means we are still far from what we would really like to know - what is the connection between energy and what many people associate with concepts such as the quality of life, sense of community, identity, esteem, happiness - sometimes also described as "soft life-values". Many social scientists are devoting increasing attention to these questions (15, 16).

It is important that this kind of relationship is more clearly identified. Inadequate knowledge of what energy really means to people gives too much scope for conscious or unconscious efforts to propagate prejudices and misconceptions concerning the various limitations and possibilities of energy policy. Despite the fact that many of these uncertainties remain, we shall proceed to a discussion of the role of energy in the economic organization of society.

Energy in the economy

Of what real significance is energy to the desires for an increasing standard of living, full employment and greater equality? This issue is one of the more hotly debated and disputed ones, partly owing to the complexity of the question but perhaps partly also because different conclusions will lead to different lines of action.

There is obviously a short-term relationship between energy supply and our total production (gross national product, GNP). Rapid and unexpected changes lead to disturbances in the social machinery. We realised that in 1973 during the oil embargo; the United States noticed it when the winter cold during 1977 became so intense that the production of natural gas was insufficient.

In the very short term the following simple model holds good:

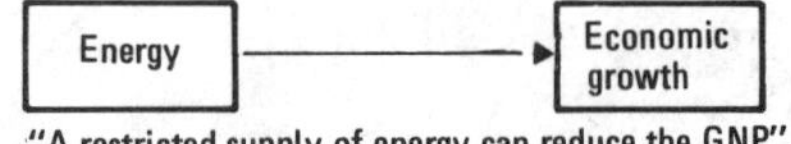

"A restricted supply of energy can reduce the GNP".

With the aid of statistical information an attempt has been made to calculate how different energy consumers react when the price of energy is raised. The figures indicate that when the energy price rises, e.g. by 10% the use of energy eventually decreases by an order of 5-10%. By and large industry seems to be more sensitive than individual households (17).

These kinds of figures - which in themselves are very uncertain - have then been used to try to analyse whether gradually rising energy prices cause any reduction in employment, economic growth and consumption.

In the long term - over a number of decades - there appear to be considerable possibilities of adaptation.

It seems that a certain degree of unanimity has begun to emerge among economists (18). If the possibilities of replacing energy with capital and labour power are sufficiently great, and provided that the technological development continues, it seems as if rising energy prices would not markedly reduce economic growth. The rise in real wages may be somewhat slower than with constant energy prices, partly because more resources have to be set aside to pay for the energy.

In the long term there need not be any connection between employment and energy consumption.

These arguments - the theoretical character of which must be emphasised - agree with those presented in other studies (e.g. CONAES (19)). When the attempt is made in connection with various model calculations to separate economic growth and energy consumption this is done by presupposing:

- more effective use of energy in individual industrial processes
- more effective use of energy by households
- that the composition of total consumption changes in the direction of more services and fewer commodities.

The "model" for the long-term relationship between energy and economic growth would presumably be that both are determined by the social pattern but hardly in any inherent way by each other.

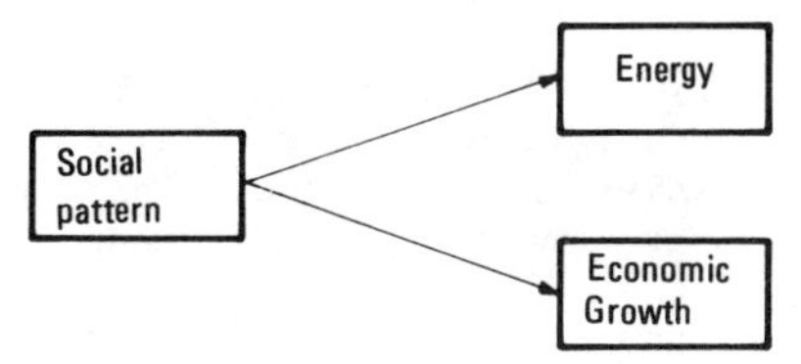

That model differs considerably from the short-term one. The question is how it looks from the middle-term perspective when energy costs will probably increase, both because situations of scarcity may arise and because new energy sources tend to provide more expensive energy than we have now.

Industry's use of energy has gradually become more effective. Seen from a longer time-perspective the changes in individual processes are fairly dramatic, as shown by the example in figure 4. It also shows that there is a thermodynamic limit in industrial processes below which one cannot fall.

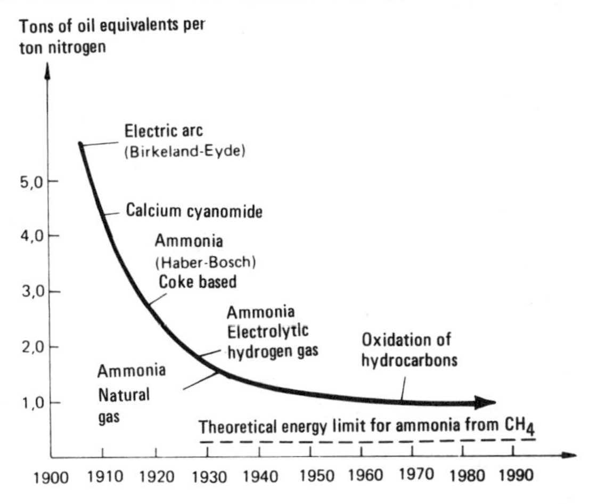

Figure 4. Energy requirements for production of ammonia by different processes.

For most branches of industry, however, these limits are a long way off - that is also true for the energy-intensive branches such as the iron and steel industry and forestry (20).

Let us look at what increasing efficiency means in practice. The cement industry is an example (21). To day there are 7 plants in Sweden with a total of 14 kilns. The energy requirement per ton of cement (kWh/ton) varies primarily according to the type of kiln:

Dry kiln	860-950
Semi-dry kiln	1100-1200
Wet kiln	1700-2100

A factory in Skovde today has the most modern dry-kiln technology and also the lowest energy consumption.

Total conversion to such dry kilns would mean that the energy requirement per ton of cement would be reduced by about a third at the same time as the number of production sites would decrease. The result would be more effective production, more effective energy use and reduced employment.

The cement industry is a very energy-demanding one but the picture is similar for others. Increasing energy costs will speed up the rate of structural transformation (22). Whether the employees can obtain new jobs or not depends on the development and direction of the economy and on employment policy.

The effect of energy saving and new technology on the future consumption of energy is illustrated in figure 5.

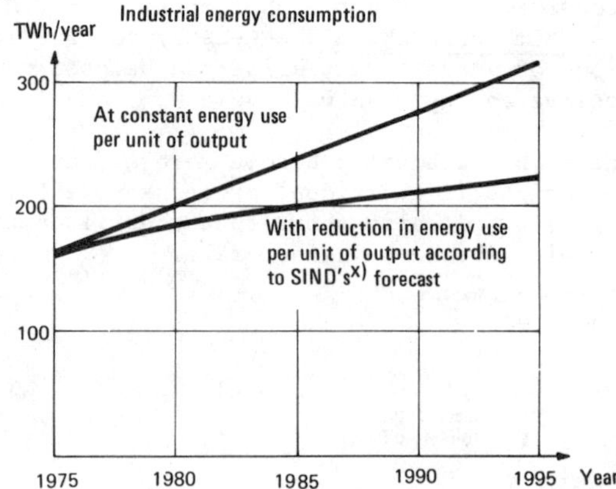

Figure 5. Comparison of industrial energy consumption with and without altered levels of energy use per unit of output. The production volume in the forecast for 1995 is 90% higher than for 1975 (23).

In the forecast of the National Industrial Board it is assumed that new technology and more effective use of energy by 1995 could reduce the energy requirement in industry by about 90 TWh. The future development of specific energy use will be primarily influenced by the technological development. How quickly changes take effect will then depend on the rate of modernisation of industrial processes and is therefore closely related to the development of the economy and the structual transformations.

(1) SIND = NIB (National Industrial Board, an agency which among other tasks has the responsibility to make forecasts of energy demand).

Leisure, housing and transport are today responsible for 70% of the energy consumption of households. (photo: Hans Hammarskiold, Tiofoto).

The problems of adaptation in commerce and industry are thus in practice a question of structural transformation.

Put another way, the question of whether or not there will be a contradiction between full employment and a reduced supply of energy (rising energy prices) depends, among other things, on the kind of employment and regional policies that Sweden is prepared to pursue.

From that perspective one can also see how much energy will be required to maintain full employment. In order to keep employment constant at a certain level it is necessary that production should increase yearly, the rate being determined by the growth in productivity. If, in addition to that, employment is to be increased, then production must increase even more, and consequently also consumption and the use of energy.

But the future energy requirement for full employment will also depend on what the national produces and consumes. If consumption is deliberately channelled into the less energy-intensive sectors of public consumption, the energy requirement will change. If 200,000 job opportunities were transferred from commodity production to the service sector the total energy requirement, for example, would be reduced by 65 TWh/year (24). The range between the energy consumption levels of different employment alternatives may be of the same magnitude as the energy savings estimated from technological measures in industry, i.e. 50-100 TWh/year. An increase in total employment by 200,000 job opportunities could mean an unchanged energy requirement, provided that the increase takes place in the service sector (24).

The adaptation to higher energy prices is also experienced in a different way by individuals and households. Higher energy prices mean that consumption is induced to develop in a less energy-demanding direction. Housing, transport and leisure are today responsible for 70% of the energy consumption of households. An adaptation to sharply increased energy prices will be reflected in housing standards, car ownership and tyspes of cars, foreign travel and other holiday habits. The adaptation directly affects our everyday lives and raises the question as to which people in the first place have to adapt themselves.

If all households overnight were to acquire the same consumption patterns as the 20% which have the highest standards, the energy requirement for private consumption in Sweden would increase by 50-100% (given the same technology). If the highest group, in addition, increases its own standards, the total requirement would of course rise still further. The higher the level at which equality is achieved the more energy will be needed.

Should the adaptation to higher energy prices take effect directly in the budgets of individual households, or should society protect those with smaller resources and simultaneously limit the consumption of those with greater resources? A distribution policy aiming at a reduction of the differences between groups could be directed in a general fashion to the distribution of incomes. But it could also be directed more selectively to the consumption of the big users.

It would, for example, be possible to alter the housing loan system in such a way that large detached houses became less attractive. One could limit the tax deductions for interest payments, raise the basic assessable income, or direct local authority planning towards other types of housing. One could restrict the tax deduction for journeys by private car to and from work. One could put the personal use of official cars on an equal footing with private cars. One could extend public transportation and tax parking-spaces at the place of work as an emolument. One could slow down the building of private weekend cottages and stimulate leasing instead. One could tax large motorboats. There are many ways of tackling the problem.

Consumption could also be directed away from private services. Health services, education and care are services which are less energy-intensive than commodity production and commodity consumption (see "The development of energy consumption" in chapter 2), and a deliberate policy of transferring resources from private consumption to those areas would also have consequences for energy policy. If everyday life included good child care, care of the elderly, medical services, education and culture, and somewhat smaller proportions of large houses, large cars, foreign travel, motorboarts etc., society would also become less energy-consuming.

Opportunities for choice thus exist and they are all related to everyday life. That is where the decisions must be taken, even if they are effected in different ways.

In conclusion we shall mention some of the areas which would be affected by an adaptation to higher energy prices. The list is not complete but should be seen principally as illustrating the fact that the possibilities in the field of energy policy are to a large extent determined by the repercussions which we are prepared to accept in other social spheres.

Industrial policy, because the changes in industrial structure and investment rate would be affected. Particularly energy-consuming plants might be eliminated.

Employment and regional policy, because the changes in industrial structure may throw many out of employment.

Capital market, because large amounts of capital will be needed for energy conservation, investment in new energy supply systems, and modernisation of industr
The increased capital requirement would mean a restriction of the scope for an increase in real wages.

Housing policy, because the costs for housing will increase, particularly with the existing stock. New housing could more easily be adapted to higher energy prices.

Traffic policy, because transport by car, lorry and aircraft will become relatively more expensive, while public transport will become cheaper. That will create a demand for an improvement of public transportation, which in turn will create demands for more finance.

Local government policy and the relation between the state and local authorities, because part of employment will (in any case) consist of health services and education - areas of local government responsibility which are largely regulated by the state.

Taxation policy, because a growing public sector, for instance, must be financed by taxes. Thereby the scope for private consumption will also be reduced.

Distribution policy, because the chances of an equalisation of standards would be restricted if the scope for increases in consumption expands more slowly.

A relatively quick adaptation to higher energy prices is not a simple process. It affects people's everyday life to a high degree and is in the last resort a matter of how people experience various aspects of that everyday life and changes in it.

Energy in the society of the future

Introduction

We have hitherto endeavoured to throw some light on the role of energy in nature, in society and also in the economic organization of society. Let us continue with a short discussion energy in the future. Why is energy a problem and why have the energy questions taken up such a large part of the political debate?

In the same way as coal replaced wood and oil replaced coal, oil must now gradually be replaced by something else. But there is an important difference between the forthcoming change of energy source and the earlier ones. This time an energy source is to be replaced by something that is more expensive. All the alternatives to conventional oil involve higher costs. That applies irrespective of whether the shift takes place to oil from shale or tar-sand, ot synthetic fuels from coal, to nuclear power or to solar energy. Practically and politically this will involve a series of problems. The desire to retain the old system will be strong, but a postponed, rapid adjustment may be more troublesome than a transition which is begun early and takes place more slowly.

Future Energy Needs....?

The degree of difficulty in adjusting the energy supply depends on the amount that will have to be adjusted. The rate of increase in energy consumption will therefor partly determine the conditions for that adjustment. Figure 6 shows the total energy consumption during the twentieth century and a number of forecasts for the period up to the turn of the century.

There is, as one can see, a not inconsiderable range between the forecasts, despite the fact that they were all made within one five-year period. The diagram illustrates the hazards of predicting future energy consumption. Let us take one forecast, EPU high, as a starting point for a discussion. Table 3 can be regarded as summarising the description of Swedish society in the year 2000 according to the alternative EPU high.

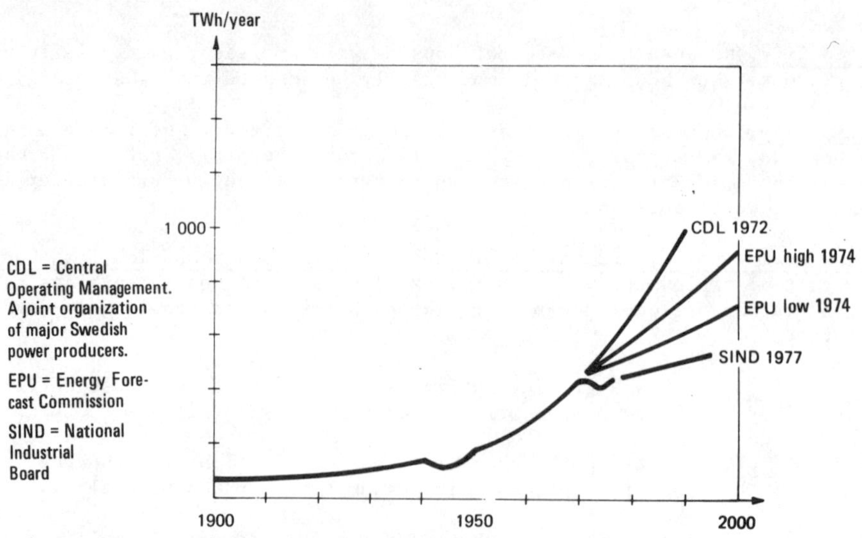

Figure 6. Energy consumption in Sweden during the years 1900-1975 and some forecasts for the years 1975-2000 (25, 26, 23).

Table 3: Production in the year 2000 in different branches and the consumption level according to EPU high alternative (26).

Branch	Production in 2000 (1970 = 100)
Mining	550
Pulp, paper	476
Chemical industry	870
Iron and metal	614
Engineering	690
Industry as a whole	575
Private cars	197
Small detached houses	146
Flats	116
Business premises	133
Schools	127
Hospitals	124

The table can be said to describe a society in which industrial production has increased fivefold, the number of cars has doubled, the rate of increase of small detached houses has been three times as large as that of blocks of flats, etc.

Naturally such projections raise a large number of questions. What are we to do with nine times as much plastic in 2000? With seven times the amount of engineering production? How will those communities function that are based on small detached houses and on private cars as the dominant means of transport?

The energy consumption in 2000 was calculated on the basis of an estimate of the production volume in different branches, the standard household consumption, and possible energy savings. The forecast showed a consumption of 910 TWh/year.

Four years later this forecast is already outdated. Not so much because the

Do we want a future society which includes twice as many private cars? (photo: Per-Olle Stackman, Tiofoto).

fundamental relationship between energy needs and production has altered, but because it is now assumed that industrial production will expand more slowly in future. Even if a more effective use of energy plays a certain role in the later forecasts of the National Industrial Board (SIND) it is the downward revision of future economic development which has had the greatest significance for the results of these estimates.

This, in turn, depends partly on the prognostic method itself - to a certain extent it presupposes the very thing which is to be proved, namely the relationship between energy needs and societal development.

Environment and Safety

Environmental and safety questions are clearly of decisive importance for future energy supplies. We have, however, refrained from presenting detailed statistical evidence. Others have already done so (27, 28) and in the time-perspective of a future study such information provides little guidance. The methodological problems in treating the data and drawing possible conclusions are complicated (29).

We shall here only deal with the issues in a very summary fashion.

Changes in the external environment - landscape changes - were the first to be taken up in Swedish politics. The expansion of hydroelectric power during the 1960's was in part slowed down on that account. Subsequently uranium extraction at Ranstad has been restricted for partly the same reasons. Other energy sources naturally also have an effect on the landscape - oil requires land for refining, coal requires land for coal - and slag-heaps, and so on. Future energy sources such as nuclear power, wind power and energy forest plantations will also have such effects.

In the combustion of oil and coal, sulphur is released. Information about sulphur is comparatively extensive. The emission of sulphur is chiefly of significance for the external environment. Through precipitation the acidity of several thousand lakes has been affected, in many cases causing the extinction of fish. The connection between sulphur emissions and acidity is well documented (30), even if the matter is still to some extent under debate (31). From a Swedish point of view it is worth noting that our country is more sensitive than many others due to the composition of our bedrock and our soils, which have a limited buffering capacity (power of resistance). It is also well documented that a large part (over half) of the acidic precipitation in Sweden derives from emissions in Europe which have been carried by atmospheric movements in over Scandinavia. The developments in this area are determined by the emissions in Europe, and even small increases there exceed the effect of large reductions in Sweden.

Other emissions connected with combusion consist of heavy metals[(1)], nitrous gases (nitric oxides), aromatic hydrocarbons, etc. These have an effect on human health. The levels of nitric oxides in the atmosphere of large cities, for instance, sometimes exceed the recommendations of the World Health Organization (WHO) and contribute to various respiratory diseases. Aromatic hydrocarbons could in theory lead to cases of lung cancer, but the uncertainties here are very great. Motor vehicle exhaust gases are now the main source of these kinds of pollution.

Among the emissions which are today relatively well known, acidification and exhau gases constitute the most acute problems. The lead emissions in exhaust gases have resulted in measurable quantities of lead in the brains of children (32), wit suspected developmental disturbances in consequence.

The effect of the energy supply on health and the environment must, however, also be compared with the influence of other factors. At this point uncertainties from two directions overlap each other and the combined picture becomes diffuse. Certa estimates have nevertheless been made for lung cancer which show that, in about 75% of lung cancer cases, smoking is a strong contributory cause, in 10-20% motor vehicle exhaust gases are strongly contributory, and all other causes (including indirect smoking) constitute the remainder.

At the present level of knowledge there is, with the exception of exhaust gases, no basis for an assertion that energy sources such as oil, coal or nuclear power play any dominant part in the morbidity of the population. Provided that effective pollution control techniques are developed and applied, e.g. to sulphur

(1) For example lead, mercury, vanadium and cadmium.

emissions, nitric oxides and heavy metals, there is no reason either, at the present level of knowledge to assert that environmental factors indicate either that we should desist from or can continue to use fossil fuels during the next few decades, but the state of knowledge is unsatisfactory. In the longer term there are several environmental problems which may emerge as more serious. This is true particularly of carbon dioxide, the possible effects of which on the climate are such that the world should not plan for an increased use of fossil fuels.

The safety problems in conjunction with disasters differ to some extent. Several elements of the present-day energy supply systems offer possibilities for disastrous events. Examples are the collapse of dams, large-scale oil fires, explosions of liquid gas stores, radioactive discharges from nuclear power stations, and so on. In most cases the chance of major disasters is considered to be very small. But even an event of a very unlikely kind could happen tomorrow. It is often difficult to assess chances of this kind with any large degree of accuracy. One is usually reduced to model descriptions without knowing how closely the model conforms to reality (33). The consequences of such a disaster could be very serious. They include many different categories of damage and inuury, immediate death, delayed death, serious illnesses, psychological troubles, pollution of land and water, etc., so that large areas might have to be abandoned for a long time. Measures to reduce the risks and consequences can be very costly. The conflicts between economy and safety can be very acute.

The overall picture of the long-term interaction between energy and environment, health and safety is thus very obscure. Uncertainty about the course of events and the consequences is considerable. There are in theory great possibilities for developing techniques and methods to deal with most of the problems, but, on the other hand, great uncertainties about whether these methods really will be developed and introduced. Certain problems in Sweden are caused by other countries, e.g. the increase in acidification, and there are doubts about what measures those countries will take. Much depends, both nationally and internationally, on how powerful the laws and organizations are which will be created to solve the problems.

An Ecológical Approach?

Ecology is another keyword in the energy debate. It is important to see ecology as something wider and deeper than the sum of different environmental questions. One often speaks of an ecological approach, of the characteristics of the eco system and the like. There are, however, many definitions. Lars Emmelin and Bo Wiman (35) see ecology primarily as a science: a natural science with its emphasis in biology. They also consider that, in questions of energy and resources, ecology is best qualified to define the framework within which all human activity must remain in order not to endanger long-term productivity. On the other hand ecology is only one of the sciences required, within that framework, to analyse and explain how the activity is to be conducted.

But ecology also means something else. Rolf Edberg (36) stated in a lecture:

"The anxiety and the feelings of homelessness which characterise our age presumably derive from the continued operation within us of primordial forces which we betrayed when we created an artificial environment. Within the new dimensions which our seeking has opened up to us we need to rediscover, in our way of life and our conduct, something of that harmony with the creative forces of nature which peoples whom we call primitive instinctively felt and translated into animistic conceptions.

Only in obedience to the laws of nature can man find his identity and thus realise himself. We must go forward - but along ancient paths".

It is not a question here of a particular collection of scientific methods or observations. Ecology (ecosophy), in Edberg's version, becomes more an all-round cultural outlook, an aspiration for a new systhesis, a way of looking at the world. Such an all-round view can be sustained by various sciences and can influence the way that various sciences formulate problems, but it is not in itself "scientific" in the current, occidental sense and can neither be proved nor refuted through research. Edberg speaks of a synthesis:

"Knowledge in itself is not enough. We need to achieve that overall grasp of our mass of knowledge which is the seal of wisdom".

Nor is a concept like ecological approach easy to translate into concrete decisions. But it does provide one of the dimensions to which scope must be allowed together with more traditional concepts in energy policy.

What is the real problem?

The very concept of energy policy is a risky one. It suggests that energy itself is the central factor in energy policy. But that cannot be the case.

If our 30 million tons of oil are a problem, it is not because they represent about 350 TWh of energy but because the use of oil causes environmental problems (sulphur, combustion products, risks of catastrophe), leads to dependency in foreign relations and is needed more by others.

If 10 million tons of coal are a problem, it is not because they produce 80 TWh of energy but because they cause environmental problems in Sweden, may cause problems of the working environment where they are extracted, make us dependent on other countries and so on.

The consequences will differ very greatly according to whether the same amount of electricity is produced in a nuclear power station, in a hydroelectric plant on the river Kalix or by wind-driven generators.

It is not the energy in energy policy which is a problem, but the consequences which different kinds of energy give rise to in relation to the environment and safety, freedom of action in foreign relations, forms of decision-making, the ecology and population structure of various parts of the country, and so on. It is these effects which have to be weighed against the value of the supply of energy

In the same way energy can never have a value in itself. It is what we use energy for that is valuable - dwelling standards, welfare, equalisation and so on.

The adjustment, therefore, has to be not between energy and economic growth or between energy and employment but between the utility of that to which the energy is applied and the negative effects of the use of the energy.

All that is really self-evident, but it is still necessary to make clear once and for all that energy is only one link among several in a complicated societal machinery.

This perspective is important for several reasons. There is no inherent value in altering energy use by one percentage or another. Nor is there any inherent value in saving energy.

An ecological perspective or two zebras which have just startled an aeroplane? (Photo: Tiofoto)

The perspective is also important for another reason. What finally determines the demands on the energy supply systems are our evaluations of various aspects of our (future) standard of life - living space, car ownership, leisure habits - as against environment, foreign dependency, centralisation, etc. The relationships may be affected over a period of time. The fundamental problem remains: how are we to shape our society from an overall point of view in such a way that life fulfils our needs, our values and our dreams?

2 Energy and Societal Development in Interaction

Introduction

A Swede in 1850 used on average about 3.5 MWh per year in the form of wood, coal or coke. Handicraft production and a guild system predominated. Industry was mainly concentrated in regions with access to raw materials, wood, coal and running water, to operate machines and processes. The population numbered 3.5 millions. About 80% of the people were occupied in agriculture and ancillary trades and 10% lived in urban communities.

A hundred years later Sweden's population had doubled. In 1950 66% lived in urban communities. The average Swede used 21 MWh per year, of which 38% was derived from coal, 28% from oil, 18% from wood and 22% from hydro power.

By 1975 the population had risen to 8 millions and the energy consumption to 56 MWh per person, of which 70% was derived from oil. More than 80% of the population lived in urban communities.

According to a forecast from 1972 the energy consumption in 1985 will amount to about 90 MWh per person (1). Another forecast (2) indicates that the energy consumption in 2000 will lie between 70 and 100 MWh per person.

The changes between 1850 and 1975 are dramatic. The forecasts of the early 70's for the rest of the century are no less striking. They well illustrate that the period 1950-1975 can be seen as unique. In this chapter we shall describe how the present position has been arrived at.

The interplay between technological, organizational and material development on the one hand and energy supply on the other is complicated, but we shall try to delineate some important aspects.

Figures 7 and 8 show the distribution of supplies of different kinds of energy from the year 1800. At the beginning of the nineteenth century wood was the dominant source of energy. It was replaced first by coal and later by oil. From the 1950's the use of oil increased rapidly and coal supplied an increasingly small share. We shall take these transitions in the energy system as a starting point of the preconditions for changes in the energy supply.

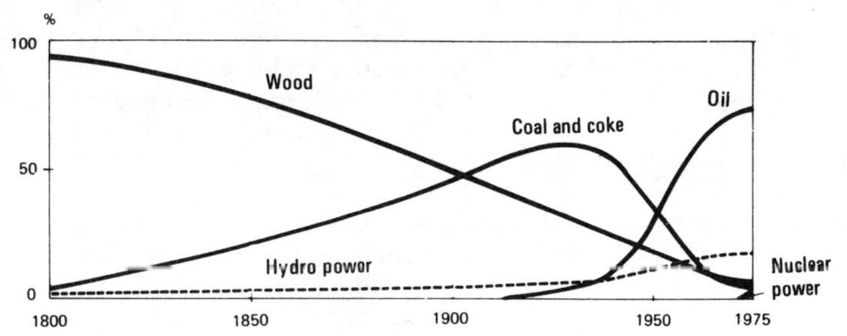

Figure 7. The share of various kinds of energy supplies in Sweden 1900-1975 (3).

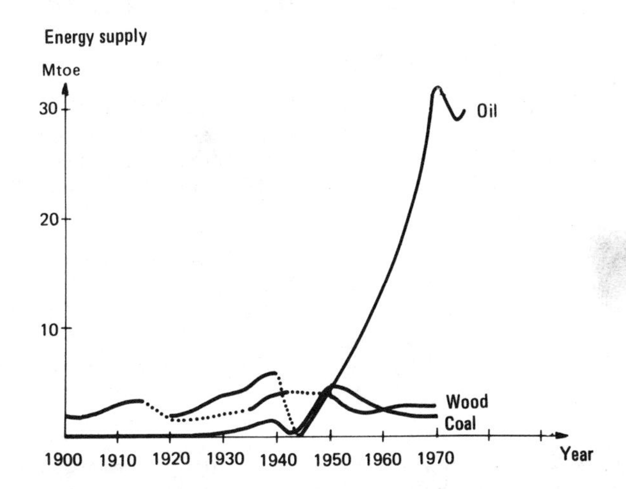

Figure 8. Use of wood, coal and oil during the twentieth century (4) expressed in equivalent quantities of oil, Mtoe.

Change of energy sources - technology and organization

As early as the eighteenth century Sweden experienced its first energy crisis. Faced with the threat of a future shortage of timber the state took steps to regulate felling and stimulate replanting. Attempts were also made to economise on wood. The government commissioned a certain Count Cronstedt to develop methods of economising on wood and one result was an improved stove. An endeavour was also made to get more people to build their houses of stone.

But an increased utilisation of timber in industry, for buildings and for energy purposes created a continuing strong pressure on the available Swedish timber resources.

In the middle of the nineteenth century timber prices rose sharply. The coal import from England, which had been started by shipping companies, increased.

At first the coal was used in the same way as wood - on open hearths and in the few steam boilers that existed in industry. But gradually new technology for the use of coal was developed. Open hearths were replaced by stoves, stoves in each room were replaced by one stove and a water-borne system of radiators. Gradually the radiators were linked together in central heating systems.

The coal also began to be used for gas production. Gas was first used for street lighting and subsequently, once that system had been developed, also in houses. There is an interesting interplay here between the growth of the gas system and that of central heating. The new capital-intensive heating system demanded cheap fuel and this was obtained in the form of coke, which was then a by-product in gas production.

In the same way as with gas the urban electricity systems were constructed around coal. Quite soon gas and electricity began to compete with each other, first for lighting, later also for cooking. In industry coal was primarily used for factory operation by means of steam engines (belt-drive, etc.)

The transition from wood to coal also involved a number of institutional changes The use of wood had to a large extent been based on self-sufficiency. It was only in towns that a sales organization existed. Coal, on the other hand, presupposed a division of labour and distributors. The introduction and establishment of coal thus became not only a question of new technology but also an organizational development. The municipalities, for instance, successively took responsibility for the delivery of gas and electricity, while houseowners, and subsequently municipalities, successively took over the tenants' responsibility for heating.

The central Swedish foundries were large forest owners. During the age of wood, industry had by and large handled its wood supplies itself. With the transition from wood to coal the coal importers assumed responsibility for the energy supply.

Last but not least, this transition led to the introduction of a whole new series of industries producing stoves, ovens and other installations for the supply and use of electricity and gas.

The transition from wood to coal involved not only the emergence of new technological systems for the distribution of energy but also new organizational forms to handle this distribution and a completely new market for coal equipment.

Now coal has long had a substantially reduced importance as energy source in Sweden. But even today we retain to a large extent the technological systems and the organizational structure that originated in the coal period.

Compared to the great changes in technology and organization caused by coal the transition from coal to oil was relatively undramatic. The first modest oil imports were for fuel for petroleum lamps. With the arrival of the motor car the market expanded so that it became meaningful to arrange for large-scale distribution. The introduction of oil to Sweden thus took place in a market where there was no real competition from other kinds of energy, namely that of motor fuel. Only later did imports of fuel oil start. Before the war it was used chiefly in houses and blocks of flats and only after the war did it come into used to any large extent in industry (cf. table 4).

Table 4. Consumption of fuel oil (thousands of m^3) (5).

	1938	1948	1951
Industry	40	1,200	2,040
Heating of premises	170	440	730

The main transition from coal to oil thus began after the second world war and applied mainly to industry and domestic heating. In both cases only small adaptations were required from the users. The systems for distribution heat in industry and blocks of flats were the same as before. The difference was that oil-fired boilers were somewhat easier to manage than coal-fired. In all its simplicity this illustrates an important basic fact: a new source of energy can expand much faster if the demands for changes to the existing equipment are small.

This rapid transition took place simultaneously with a substantial increase in the total energy consumption. From a technological point of view nothing much really happened. The water-borne systems expanded and a gradual change took place from heating of individual properties to district heating. The expansion in motoring was favoured by the fact that town planning and road construction were designed with motoring in mind.

It is also worth noting that the use of oil during this period increased most rapidly in countries which, like Sweden, lacked coal resources of their own. The chain of production from oil well to oil user was controlled, then as now, by a small number of very large enterprises which, in their turn, maintained a widely ramified system of collaboration (6, 7). The link between coal and coal user hand been more fragmented - there was no equivalent either of the integrated organization of the oil companies - embracing the whole chain of production from oil well to petrol station - or of the close collaboration of the oil companies with each other along the whole of that chain. Coal had been imported to Sweden by ship-owners and general merchants and had been resold to wholesale companies and retailers. Before the second world war there were around 40 general merchants, about 145 wholesale companies and maybe 5,000 retailers. The organizational fragmentation of the coal industry did not make it easier for it to compete against the oil companies.

The rapid increase in the use of oil is obviously explained by the low prices, its many uses and ease of handling. The oil companies had since the 20's controlled production, above all in the oil-rich Arab countries, through a number of joint production arrangements. The governments of the oil countries were certainly weak, but they kept pressing for increased revenue from their oil. The American market had been closed to imports by import quotas and the companies were virtually forced to find an outlet for the large production from the Middle East. In the background was the threat that independent oil companies (e.g. ENI in Italy) would get access to Middle East oil. It was in the interest of the oil companies to increase the use of oil rapidly in countries where that was possible.

In step with the increased imports of oil the responsibility for the Swedish energy supplies gradually passed over to the international oil industry. The movement towards a growing dependency on other countries, which had started in the coal period, was reinforced. Capital intensity and specialisation increased.

In the same way as coal, oil gave rise to new branches of industry, above all in the transport sector. The extensive collaboration between the oil companies and the automobile industry has been documented in various contexts (8). The two are like Siamese twins.

The Swedish electricity system

After this short historical outline of different fuels, let us proceed to the development of the electricity system. The Swedish electricity system grew up around two "growth centres". One was the Swedish towns, which had established themselves through their gasworks as producers and distributors of energy, the other was the iron-working and paper industries in central Sweden which had large forest resources and thereby riparian rights to a number of exploitable waterfalls. The towns also began to utilise coal to produce electricity and, in addition, expanded their own distribution system. The processing industry began early to invest in hydroelectric power for itself. As a surplus of electricity often existed the power plant owners stimulated the formation of local distributive associations in the surrounding areas to which the surplus was sold.

Once the technical difficulties of long-distance transmission of electricity (alternating current and transformation) had been solved, it was no longer technical difficulties which restricted the harnessing of hydroelectric power but institutional ones. The Water Rights Act, for instance, required that all riparian owners should agree on construction of a power station. Many legal measures were required before hydroelectric power could be exploited (9).

By a new Water Rights Act of 1918 many obstacles to the harnessing hydroelectric power were removed and for a long time it was subsequently limited mainly by the development of the market. The power producers therefore took the initiative in stimulating the distribution system. By the mid-1930's Sweden was by and large electrified and the Swedish electric power organization had taken shape. A relic from the coal era was that distribution and production were organizationally separated. Production was divided roughly in equal parts between The State Power Board and a number of municipal and private countries. The latter were owned in their turn by the large electricity consuming industries. The distribution was organized by local authorities or by distributor associations, above all in the countryside.

During this stage it is clear how strong the link between producer and market was. It was the towns which replaced gas lighting in the streets with electricity - when gas was already established for domestic use - and it was the electricity-using industry which first developed hydroelectric power. The surplus(1) in industry provided a stimulus for the development of the distribution system for domestic consumers. The same applies to the State Power Board. The power station at Porjus was originally constructed for the electrification of the railway connecting Kiruna and Narvik over which the iron-ore was transported, but subsequently produced a surplus for many years which, in its turn, gave rise to a wider distribution.

In the mid-1930's the development strategy of the electricity producers switched from increased connections to increased volume. The rates were changed - the earlier low, fixed charges were raised, while the energy price was lowered - and a fairly intensive marketing of electricity-using apparatus began. Organizations were formed to stimulate an increased consumption of electricity (e.g. FERA - the Association for the Rational Use of Electricity).

The next phase involved the successive linking up of the regional electricity systems through the main transmission grid. At the end of the 1950's the electricity system consisted of an established group of distributors and an established and increasingly collaborating group of power producers.

During the 1950's nuclear power made its first appearance - the first electricity-producing installation started in 1954 in Britain - and at the same time the price was gradually lowered. The opposition to a further extension of hydroelectric power increased and the electricity producers looked around for other opportunities for electricity generation. One possibility was nuclear power, the other was oil-based condensing power stations. Both could in theory be designed as combined fuel and power installations, i.e. for the production of both electricity and hot water. Nuclear power was advocated by the electric power industry (ASEA) and the State Power Board, oil-based heating and power stations by the local authorities, which then saw a chance of combining district heating with electricity production

(1) The surplus results from the difference in time between the electricity demands of industry (daytime) and domestic consumers (morning, evening). The productive system therefore did not need to be extended to meet the electricity needs of domestic consumers as well.

Electricity has started to make its entry into Stockholm. At Slussen an electric tram drives past the high lamp-posts with their bright arc-lamps. (Photo: Stockholm energiverk).

Power station construction at Porjus around 1912. The development of hydroelectric power became an exponent for the industrial society which grew up at the cost of the old agricultural community (Photo: Vattenfall).

for their own distribution.

But a rapid expansion of co-generation plants run by local authorities constituted a threat to the power producers in general and to the State Power Board in particular. Towards the end of the 1950's the State Power Board was in fact in a difficult situation. There was surplus capacity, and sales to industry were stagnant. At the same time low-voltage deliveries, above all to domestic users, were seen to be expanding rapidly, particularly in the growing urban areas. That was where the future market for electricity lay. But between the State Power Board and the millions of low-voltage subscribers were the distributors - e.g. the municipal electricity works.

At the beginning of the 1960's a large number of cities were preparing to launch into co-generation plants.

This would have three consequences for the State Power Board:

1. the market for electrical heating would be restricted as district heating was an alternative.

2. the growing markets for electricity production would in the first instance be supplied by the local authorities' own co-generation plants. Thereby the market for future nuclear power would also decrease.

3. the role of the State Power Board would change from that of the dominant power producer to that of equaliser, producer of standby power etc., whose conditions were decided by others.

The cities were not otherwise independent of the electricity system. The co-generation plants could not operate in isolation. They had to be connected to the grid in order to have access to standby power or to dispose of surplusses. Vasteras was the first town that managed to build its own co-generation plant by associating itself with one of the smaller power producers, Bergslagens Gemensamma Kraftforvaltning (BGK) (10). The cities of central Sweden also tried to set up a cooperative organization of their own, AB Mellankraft, in which surplus capacity from a co-generation plant in one town could be sold to the electricity works of another town.

The State Power Board, as largest power producer and administrator of the main transmission grid, was in a position to influence the economy of the co-generation plants. The counter moves of the State Power Board were also chiefly economic. The rates for large city areas were lowered in 1963 and 1965 and contracts were signed with the municipal electricity works of such a nature that the latter did not find it profitable to build power and heating plants. The contract periods covered 10-15 years. The cities which were threatening to produce power and heating plants were thus given a discount on electric power in return for undertakings not to generate power themselves. The State Power Board also demanded that if a municipality wished to terminate the contract and build a co-generation plant, the State Power Board should be called upon to investigate the project. The cost of standby power was high, at the same time as the payment for surplus power was low. The result was not long in coming. The number of cities which established themselves as co-generation producers stagnated during the 1960's. This threat to the future of nuclear power was thus averted. The duel between co-generation and nuclear power is clearly shown by the following report of a debate during this period.

The duel between co-generation and nuclear power

"The co-generation plants were undoubtedly beginning to play a certain role in the energy market. With the Adam project in Vasteras one may say that an atomic age which left its mark in the debates on energy supplies was initiated. The attitude of the State Power Board to co-generation power became more and more negative. At the meeting of the Association of Electricity Supply Undertakings in Orebro in May 1965 - two days after the inauguration of Orebro's municipal co-generation plant - it manifested itself in a duel between two trends of thought; one, designated co-generation, represented by the director of the Orebro Industrial Board, Olof Blomqwist, and the other, designated atomic power, represented by Erik Grafstrom, Director General of the State Power Board.

Blomqwist held that electricity supply is already moving from the epoch of hydroelectric power to the age of co-generation. He pointed out that complete large-scale power and heating stations are today operational or under construction in eight Swedish towns. Of these at least the towns in the Malar Valley are planning cooperation in this respect. He found it not unrealistic to think that towns with as few as 20,000 inhabitants would construct power and heating stations in the future. One of his conclusions was that there was reason to doubt the urgency of atomic power.

This provoked Grafstrom to a forceful reply. He feared that if the production of power and heating stations and hydroelectric power increased to the extent indicated by Blomqwist, our country would not be able to participate in the development, which "is now storming through the world", towards large-scale stomic power installations. Grafstrom also feared that the questions of cooperation would be even further complicated by increased fragmentation on the side of the power producers, as in the case of retail distribution.

That was a broadside, delivered by a man in battle array, even if it was a little difficult to understand that there was any cause for it. On the same occasion another of the State Power Board's men explained in conversation that a thermal capacity must be developed in the field of electrical heating in order to be able to utilise the atomic power which was apparently going to flow in abundance. It was hard to grasp how the situation could be handled until the flow began.

The attitude to co-generation had changed from the time when it was decided to test the first atomic-driven thermal reactor on the basis of Vasteras' thermal capacity (10)."

Some observations

Oil expanded very rapidly in Sweden - it took 25 years to rise from a few million tons to the roughly 30 million tons today. It has, however, taken longer to introduce other energy sources - coal, hydroelectric power, nuclear power. It took 20 years to develop legislation so that hydroelectric power could expand. It took 20 years from the initial plans for nuclear power until nuclear power became a noticeable factor in the Swedish energy balance.

The change-over involves not only a technological change in which one kind of equipment is exchanged for another, but also a number of economic and institutional changes. Old forms of institutions must be replaced by new ones better adapted to handle the new techniques.

There is here a complicated interplay in which, for a time, technological conditions and limitations determine the economy and organizational structure. New technology must usually be compatible with the old. To close down functioning installations is often expensive. Adaptation of new technology so that it can function together with the existing techniques - like a jigsaw puzzle - makes it possible to begin in a small way and then gradually expand. The adaptation also includes control of the conditions for connecting up of new plants, legislation, finance, corporational forms, and so on.

In the following section we shall therefore discuss how the use of energy has developed and try to describe the interaction between social development and energy consumption.

The development of energy consumption

Wood-Sweden was to a large extent an agrarian society. The wood was used for domestic trequirements and only a small part was sold. The transport costs were at that time so high that industrial development was impossible except where the combination of wood and other raw materials was uncommonly favourable (e.g. in the Bergslagen mining area). The introduction of coal, therefore, involved not only a new fuel but also a fuel with such properties that industrial development became possible. Production could be concentrated without the transport costs for energy becoming unreasonable. The advantages of large-scale operation could be utilised and the social organization gradually changed. The increasing division of labour was both made possible by and presupposed increased production. Oil and electricity which came into growing use in the 30's, made new ways of life easier. Comfort increased, the time spent in dealing with heating and various domestic tasks decreased.

The increase in oil consumption since the 50's has become necessary, among other reasons, in order to cope with greater dispersal of the population and of community services. Changes in the patterns of living, for instance, have implied changes in the forms of service which we require. The family and relatives have in all periods had a large share of the responsibility for the care of children and old people. This has become more and more difficult to maintain when distances are great and people have increasingly come to participate in economic activities outside the home. The solution was first hired labour within the family, then day nurseries, hospitals, homes for the aged. The demands for buildings and transport increased. Despite greater comfort people had less and less time. In any case the consequences with regard to energy were demands for rapid transport and time-saving domestic appliances. An conversely - these changes could hardly have been effected without increased energy consumption.

The development can be summarised in the following model, where H represents housing, W work, S service and L leisure (11).

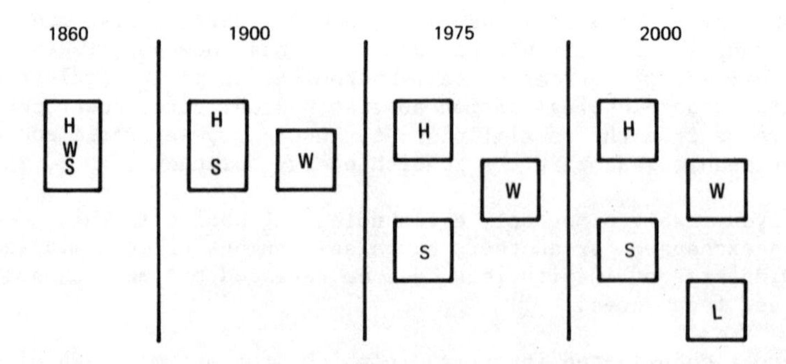

We shall illustrate the development in the postwar period in somewhat greater detail.

It is both easy and difficult to describe what has happened since, say, 1950. It is easy if one contents oneself with traditional descriptions such as gross national product (GNP) and energy consumption. In that case the development can be summarised in the sentent:

"When GNP increased by 1% energy consumption increased by almost 1.5%"

Sweden's economy has during the last few decades become even more energy-intensive. It is when we try to explain why this has happened and what the increased energy consumption has involved that the difficulty begins. Between 1960 and 1970 energy consumption doubled - from almost 200 to about 400 TWh per year.

Two areas dominate our energy consumption - private consumption and exports. The share of the former is about 50%, although we have not included in this the imported energy which is locked up in commodities. That amount of energy has been estimated for 1970 at about 100 TWh.

In chapter 1 we established that housing, transport and leisure together constituted about 70% of domestic energy consumption (in 1968). In what way has energy consumption increased?

Dwellings have become larger, warmer and more numerous, and Figure 9 shows that the energy consumption for dwellings has almost quadrupled.

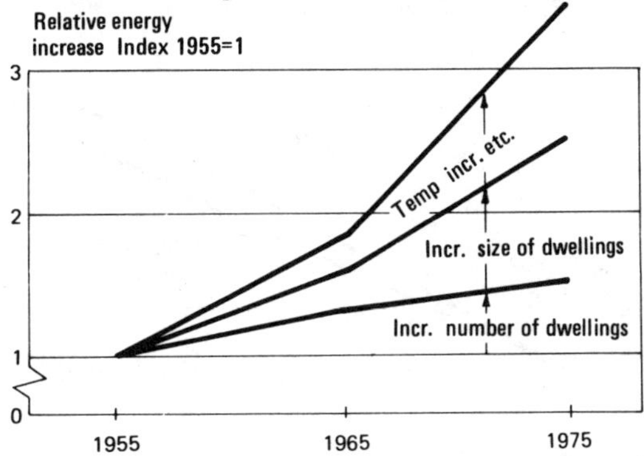

Figure 9. Estimate of the effect of various factors on relative increase in energy consumption for dwellings since 1955 in Sweden (12).

An improved standard of housing has also led to increasing domestic equipment. Table 5 shows the development of the supply of various electrical appliances in households.

Cars have also become more numerous and larger. The annual driving distance, on the other hand, has not changed significantly. Figure 10 shows the increase in energy consumption for private cars, which dominate passenger transport (86% of the energy consumption for passenger transport and 48% of the total transport sector in 1976 (12).

The number of charter flights abroad has increased tenfold between 1960 and 1975 and the number of weekend cottages has nearly tripled during the same period (15).

Table 5. Share of households having various electrical appliances (in %) and the average energy consumption per year for one appliance (13).

	1964	1969	1973	Average consumption/ kWh/year
Electrical cooker, hotplate	70	83	87	830
Refrigerator	75	92	94	630
Freezer	22	46	60	700
Washing machine	29	41	49	400
Airing cupboard	-	-	9	700
Dishwasher	1	5	11	370
Vacuum cleaner	84	89	90	100
Hair dryer	20	43	61	100
Sauna	-	2	4	470
TV, b/w	76	83	70	230
TV, colour	-	-	27	300

Larger dwellings, private means of transport, simplified domestic work and increased leisure activities have for large groups been the most important part of the rise in standards. The increased energy consumption is therefore partly a result of the fact that a living standard which just after the war was reserved for a minority has now become that of the majority of the people. At the same time the energy consumption of the well-to-do has continued to increase - this also concerns the size of dwellings, number of cars, travelling, pleasure craft and weekend cottages.

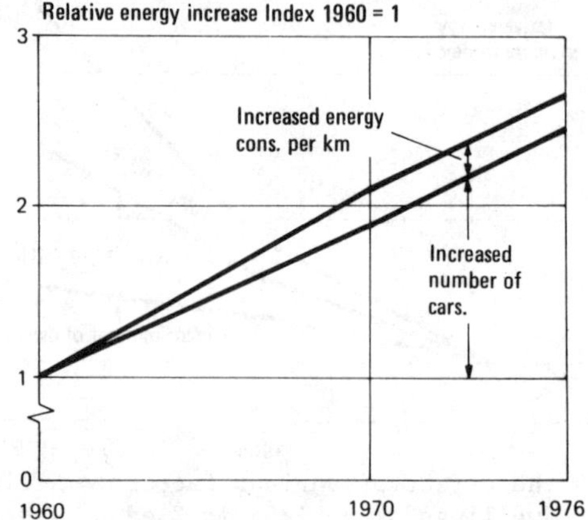

Figure 10. Estimate of the effect of various factors on the relative increase in energy consumption for private cars (14) (the driving distance per car has remained unchanged).

It is important to establish that these changes have not resulted only from the free choice of consumption and the increasing economic resources. In several respects the development has been promoted and stimulated by the state through direct measures or as indirect effects of the other measures and decisions. A few examples may be appropriate.

The expansion of motoring has been promoted by the fact that community planning since the early 50's, when the number of cars was still limited, presupposed more cars (17).

Motoring has also been stimulated by the opportunities offered by the tax system for deduction for journeys to and from work. The car as an emolument is not taxed in the same was as salaries, so that more and more companies have discovered the advantages in providing employees with a car rather than raising their salaries.

An increasing number of household capital goods have been included in the terms of the cheap housing loans. Dishwashing machines are today a borderline case - the space for a machine is today eligible as security for a loan but not the machine itself.

The increasing domestic energy consumption has been stimulated in the same way through housing policies. Cheap state loans together with housing allowances have enabled many households to have considerably more spacious accommodation than would otherwise have been impossible. In addition, the combination of high marginal taxes, the unlimited right to deductions and an accelerating inflation have raised the interest of many high and medium income earners in living in detached houses which - in turn, due to the size of the accommodation and longer transport - has pushed up energy consumption.

The removal of the restriction on loans for houses larger than 125 m^2 has, together with inflation, contributed to the increased dwelling space in detached houses. The rules of the capital market have made it possible for many houseowners to extend their mortgage and their mortgage and thereby purchase a weekend cottage, to mortgage that, purchase a pleasurecraft, and so on.

An analysis of the development of Swedish energy consumption shows that industrial production has become ever more energy-efficient, reckoning by units produced. At the same time consumption has become every more energy-demanding.

Table 6 shows that private consumption absorbs ever more energy per unit and at the same time provides less and less employment.

We may here refer back to the argument in chapter 1 that industrial production in reality consumes energy in two stages - firstly during manufacture itself, secondly in use of the products. The development can thus be summarised as follows: the products of industry consume a decreasing share of energy during manufacture and an increasing share of energy when they are used.

Table 6. Energy and labour power requirements for private consumption (18).

	kWh/Kr	Working hours/Kr
1960	2.1	0.062
1965	2.6	0.047
1970	3.1	0.026

Half of the energy used by domestic consumers was direct (i.e. for running vehicles, appliances, etc) in 1960. By 1970 the direct share had increased to about two thirds (13). What can be the reason for that? Let us try to see the connections between the energy requirement per unit produced, the energy requirement per employee and production per employee, i.e. the productivity of labour. These three factors are obiously interrelated but we shall start by investigating the relation between productivity of labour and energy requirement per unit produced in some different fields of activity.

In industry the increased productivity of labour has not involved increased energy consumption per unit produced. Energy-saving techniques have compensated more that well for the introduction of more machines.

In the transport sector the productivity of labour has increased, but so has the energy consumption per unit produced. This is due to reduced transport times, i.e. speeds have increased. Time is bought with energy

In the forestry and agriculture sector the productivity of labour has also been raised by means of a strongly increased energy input in the form of more tractors, more drivers, more artifical fertilizer, etc. In broad terms energy from people and animals (horses) has been replaced by machines, at the same time as the yield per person has increased.

For the service sector, finally, there are not very many statistics, but what there is indicates that here as well the energy requirement per unit produced has increased together with the productivity of labour. One probable cause among several is the larger size of premises. We surround ourselves with more and more floor space in offices, hospitals, shops etc.

In industry there has thus been a strong increase in productivity at the same time as the energy requirement for the manufacture of a product has not increased. But how has the energy requirement per employee changed? The development between 1960 and 1970 is summarised in Figure 11, which shows how the energy requirement per working hour has changed for different purposes.

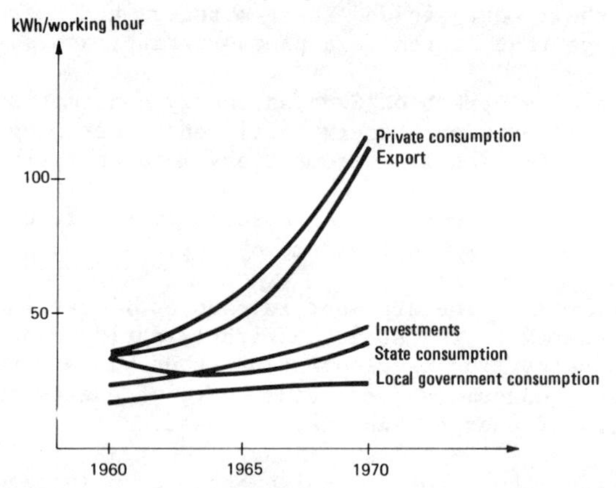

Figure 11. Energy requirement per working hour (average values) within different sectors. The values have been obtained by dividing the energy consumption within a sector by the number of working hours within the same sector (18).

An individual working in 1970 in production for private consumption involved the use of almost five times as much energy as if the same individual had worked in the local government sector. The most striking fact is that the differences have increased so sharply during a period of 10 years.

As the energy requirement per unit produced did not increase in industry, the change must be attributed to the increased productivity per employee.

Contributory factors in the postwar development are the price trends for energy, labour and capital. Both energy and capital have become relatively cheaper than labour. International comparisons (19) also show that the relative prices have a clear significance for the choice of technology for various industrial processes. The same applies to households. Energy has, like domestic appliances, cars and housing, become relatively cheaper than services and commodities where it has not been possible to reduce the labour input to the same extent (13).

It is only recently that the price trend has been reversed. The price of oil rose slowly in the early 1970's and explosively during the oil crisis. Since then, however, inflation has reduced the impact of the rise in price of oil.

We end this chapter with a discussion of how consumption and supply of energy interlock in politics.

Energy in politics

Energy was as important to Sweden in the eighteenth or nineteenth century as it is today. Its importance is not determined by the amount of consumption but by the role that it plays in the country. The state has always been concerned with an responsible for the functioning of the supply of energy. During the eighteenth century the government took the initiative in methods of saving wood, and one of the results was an improved stove. During the whole of the nineteenth century one finds the governors of various provinces worrying about the supply of wood. The willow avenues in Scania, for example, with annual growths of shoots, were planted in order to meet the need for fuel and winter fodder (20, 21).

Energy supplies have always been linked to the standard of living and prosperity. The attitude to the supply of energy has, in that sense, generally remained the same. On the other hand the ideas as to which mechanisms should guarantee the supply have varied. One disputed question has been whether the government should involve itself in conservation or whether its responsibility should be limited to creating a favourable climate for energy producers.

The debate about economising in energy is not new, not even during the twentieth century. Thus the 1951 Fuel Commission, in preliminary reports, proposed active efforts to save energy as the best way of minimising the threat of oil dependency (22). The Commission recommended, for instance, the insulation of houses, the development of wind power and heat pumps.

But when the Commission presented its final report in 1956 no thoughts remained about an active policy of economy (23). The import dependence was to be solved not through abstinence on the part of users, but by taking measures to supply energy chiefly by means of domestic nuclear power.

One can only speculate about why the attitude changed between 1951 and 1956. One reason was undoubtedly that domestic nuclear power made its appearance as an alternative to oil. Production in the large oil fields in the Middle East was also increasing strongly, which led to gradually falling oil prices.

Another reason was probably that a policy of conservation is administratively more complicated than a policy for increased supply. In addition, faith in the market economy grew during the 1950's and administrative interference and control was deprecated (17). Swedish energy policies during the twentieth century, as expressed in government publications, enquiries, parliamentary bills etc., have very largely been electricity policies, although Swedish energy consumption in chiefly a matter of fuel. This is an important observation. Certain energy systems presuppose that legislation etc., is adapted to the demands of new technology, while other systems do not. Oil could replace coal by and large within the framework of the existing body of regulations. Electricity in general and hydro power in particular, however, presupposed new legislation on such matters as monopolies and concessions, the rights-of-way of power lines, requisitioning of land, and safety regulations.

It is presumably for that reason that there has always been a dispute about the role of the state as energy producer. That was true as early as the beginning of the century, when it was debated whether the state should engage in the harnessing of hydro power or should leave that to private entrepreneurs. Particularly the conservatives at the time were divided on that issue. In a similar manner there was a general political consensus about the introduction of nuclear power in the 50's and 60's, but on the other hand a dispute as to whether the nuclear power industry should be nationalised or private. The solution was mixed nationalised and private enterprises.

The formulation of fuel policies after the second world war is also connected with attitudes to the role of the state. The 1946 Commission on the Oil Trade was very critical to the international oil companies and devoted much effort to documenting various endeavours to form cartels (24). The Commission proposed that the oil trade should be nationalised. But nothing came of it and the next commission to discuss the question, the 1951 Fuel Commission, assigned to the trade association of the oil companies themselves the task of describing the oil trade (5). In the Commission's final report (23) a further step was taken and it was now stated - without explanation - that the international oil companies not only felt a responsibility for the Swedish market but were also engaged in "life-and-death" competition:

"The big oil companies hare highly expansive and motivated by a determined wish to spread oil over as large areas as possible. The competition between them is intense and large oil customers have - so far at least - been able to negotiate favourable prices. That the oil companies which are active in Sweden do have a sense of responsibility for the provision of the Swedish market was shown by the events in connection with the oil crisis in Iran a few years ago. Western Europe - including Sweden - was then largely supplied with Persian oil through British and Anglo-Dutch oil companies. When the Persian oil was stopped, an extremely critical situation therefore arose. Deliverance came in the form of exemplary collaboration between the oil companies, which were otherwise engaged in life-and-death competition. Cargoes afloat were redirected throughout the world and the shipments crossing the Atlantic changed course. With good cause the saying was coined that the Persian oil crisis affected the whole world - except the consumers". (25).

The oil companies are in a position of much greater freedom in Sweden than in most other European countries, according to a study by the United States Federal Energy Administration (26).

One can roughly divide decisions on energy policy into those which influence the structure of energy supplies, and are thereby in someway innovative, and those which are more in the nature of supplementary decisions to the innovative ones. A list of the former decisions is as follows:

- the creation of the State Power Board in 1909
- the new Water Rights Act 1918-20
- the nationalisation of the main transmission grid in 1946
- the decision not to nationalise oil supplies but to accept the international oil companies (1950's)
- the introduction of nuclear power 1955-60

The state has in the main accepted the institutional structure demanded by the technology. The way has been smoothed, but the direction has hardly been determined.

Apart from these major decisions, however, the state has introduced relatively extensive and detailed regulation of various aspects of the energy supply. If one examines its various actions, the image of the role of the state becomes contradictory. We can, for instance, identify the following roles which have successively been evolved and distributed between different organs of the state (companies, agencies, the government, etc.)

The state is a producer of energy through the State Power Board and also has certain semi- or completely nationalised enterprises on the production side (ASEA-Atom, Uddcomb, OPAB, Svenska Petroleum, and others).

The state is a consumer of energy in government buildings and defence and is therefore also a purchaser.

The state regulates the activities of the energy enterprises in order to protect the interior and exterior environment, questions of military preparedness etc.

The state directly and indirectly affects the financial situation of the energy enterprises through the fact that the State Power Board's investments are part of the national budget, that the Bank of Sweden gives preference to power enterprises on the capital market, that the state regulates (or refrains from regulating) the principles of tariff-fixing and internal financing, etc.

In additional the government is the court of highest instance in a number of legislative processes which concern energy management,e.g. the Atomic Energy Act, Water Rights Act, Building Act.

A division of roles has thus gradually emerged in which those with direct responsibility for energy supplies comprise companies and organizations with very different terms and conditions - state-owned corporations, local government bodies, trading partnerships, multinational oil companies, etc.

The public system of regulations may be compared to a passive filter in which plans and proposals from the energy enterprises are accepted, modified or rejected. On the other hand the state has not had an active and impelling function apart from funding such research and development as the energy enterprises involved do not wish to finance themselves.

Conclusions

In this chapter we have tried to show the web of connections that exists between supply and consumption of energy. From an economic point of view the national economy has become more and more energy demanding. Consumption has gradually changed in an increasingly energy demanding direction.

Seen from a technological point of view energy supplies have changed from solid fuels to liquid fuels and electricity. These changes have taken place gradually over decades and the energy prices have been falling.

On this technological and economic point of view one can then superimpose an organizational and a legal one. The state has subordinated energy supplies to the demand for economic growth and in addition (consciously or unconsciously) has designed the system of regulations in such a way as to stimulate energy demanding consumption. In relation to the energy enterprises the state has in all essentials acted passively - made adjustments from the points of view of environment, defence and the like, but in the main left future planning to the enterprises.

It is also clearly evident that the introduction of new energy sources is a lengthy process. A long series of gradual adaptations of technology, corporational forms, and legislation, must often take place before a new technology is established. This becomes most apparent when one looks at the development and introduction of, first, hydroelectric power and, then nuclear power. The lapse of time has here been of the order of several decades between the idea and significant additions of energy.

3 Sweden's Energy Supply from an International Perspective

Introduction

Any discussion of future Swedish energy policy must begin with analyses of the international development. Sweden today imports about 4 tons of oil per inhabitant and year. That is the highest ratio in the world. Oil represents about 70% of our energy supplies. If we are to utilise coal or natural gas we must also be able to import. For nuclear power, too, we are dependent on foreign services with regard to the enrichment of uranium and possibly the reprocessing spent fuel. Today we also import the uranium.

In this chapter we shall try as briefly as possible to discuss the world energy supplies and Sweden's chances of gaining access to oil, coal, gas and uranium.

The problems have been more thoroughly discussed in some subsidiary studies within the project. We refer interested readers to the reports "On oil supplies", "On the nuclear fuel cycle" and "Understanding coal". (1)

Discussion of the world energy supplies has two aspects. One concerns how energy consumption develops. The other deals with different energy sources, their availability and usability. We shall start with energy consumption.

Global energy consumption

In 1975 the world turnover of energy amounted to the equivalent of over 6 thousand million tons of oil. How was all that energy used?

We have previously seen that the Swedes use varying amounts of energy depending on their economic living standard. The same naturally applies to the world as a whole. The differences between countries are very large and this reflects mainly differences in economic development but also differences between the various countries' ways of using energy. Sweden, for instance, has the same income per inhabitant as the United States, but we only use half as much energy per inhabitant (1). A few examples are given in table 7.

Many forecasts have been made of future energy requirements in the world. Most of them extend towards the turn of the century. A few have a long time perspective

(1) In Swedish only

Table 7. Energy consumption etc., in some of the richest and poorest countries in the world 1973.

Country	Total energy consump- tion TWh	Share of global consump- tion %	Energy consump- tion per inhabitant MWh/year	Income per inhabitant (5) Kr/year
United States (2)	19300	33	92	26300
Japan (2)	3280	4.7	31	13300
German Federal Republic (2)	2860	4.5	46	20000
Great Britain (2)	2400	4.1	43	13000
Brazil (2)	466	0.8	5	1700
Tanzania (4)	110*	0.2	8	500
India (4)	1670**	2.7	3	-
Sweden	428	0.6	53	24000

* of which 92% non-commercial energy (wood, refuse, dung, etc)
** of which 70% non-commercial energy (wood, refuse, dung, etc)

Most commonly the starting point has been forecasts of future economic growth, after which the energy requirements have been calculated using more or less advanced mathematical models. The study that was commissioned by the World Energy Conference (WEC) (6) is a typical example. Its conclusion is shown in the following diagram (Figure 12).

According to the WEC's study the energy requirement in the year 2000 will be between three and four times larger than today. It is based on the assumption that economic growth will lie between 3.0 and 4.1% per year and that energy will be used more effectively. The study has been criticised, e.g. from India, for not setting up the aim of a more equitable distribution.

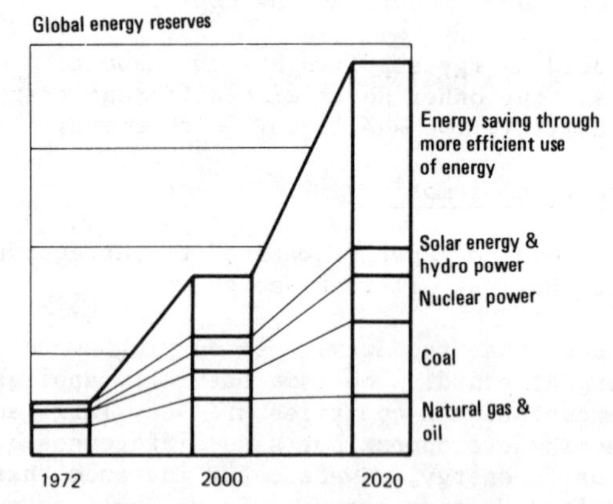

Figure 12. Global energy reserves according to a study for the World Energy Conference (WEC) (7).

A study from IIASA had a somewhat different starting point (8). Here the energy demand was calculated with regard to the population growth and assumptions of energy consumption per individual. Table 8 gives a few examples of how much energy may be required in the year 2030 for a world population estimated to be 8 thousand millions (twice as large as today).

Table 8. Energy requirement in the world at different levels of average requirement per inhabitant and a population of 8 thousand millions in the year 2030 (8). The present global mean value is about 16 MWh/inhabitant and in Sweden 53 MWh/inhabitant.

Global mean value	Global energy requirements
18 MWh/inhabitant	about twice that of today
27 MWh/inhabitant	about 3 times that of today
45 MWh/inhabitant	about 5 times that of today

A rise in the energy demand of the developing countries to a level near that of Sweden today would increase global energy demand 5 times. If, in addition, the energy consumption of the US and other industrialised countries rises, the total requirement will obviously rise still further.

Studies from industrialised countries,with their starting point in assumptions about economic growth,and studies which primarily focus on the development of the developing countries (though in the foot-steps of the industrialised countries) thus both point to greatly increasing energy demand.

Energy supplies - in industrialised and developing countries

It is not possible here to continue the discussion of the development of energy consumption. We shall instead proceed to energy supplies, but will remain for a while at a developing country perspective.

In Sweden, due to our high technological level, we can discuss different kinds of energy sources (oil, gas, electricity, district heating) as relatively comparable. From the point of view of the developing countries they are in many cases not at all comparable. Some energy sources demand a well developed organization to make their utilisation possible (electricity networks, gas pipelines etc.). The problem of many developing countries is precisely their lack of such facilities, which they also have insufficient capital to construct them. For that reason they are unable to exploit certain kinds of energy.

Oil is in many ways an exception. It is easy to transport and store and has a high energy density, while it can also be used as a fuel.

If the developing countries do not get access to cheap oil their chances of experiencing a development like that of the industrialised countries are seriously limited (3). Their alternative in the short run is to continue to use non-commercial energy such as wood, dung and so on. The ecological damage caused by continued overexploitation is already great in certain areas and may be further aggravated (9). But in the not too distant future the developing countries should have good prospects of utilising renewable energy techniques (10, 11, 12, 13).

The fact is, however, that the oil at present used by the industrialised countries is to a large extent (about 80%) located in developing countries, while for instance coal and uranium, which the industrialised countries may depend on in future, are chiefly located in the industrialised countries; the share of the industrialised countries can be estimated at roughly 75% of the total known reserves (3).

The dependence of the developing on the industrialised countries may therefore increase in step with the exhaustion of the developing countries' oil by the latter.

Oil

In 1975 2,700 million tons of oil were consumed in the world as a whole, which represents about 44% of the total energy consumption. Sweden's oil imports were expected in 1975 to rise from about 30 to 33 Mtons in 1985 (14). The forecasts of SIND[1] also point to somewhat higher oil imports than today (34). Other countries as well expect an increase in oil consumption - it is estimated that consumption in the OECD countries will rise from 1800 Mtons in 1974 to 2500 Mtons in 1985 (15). The developing countries are planning for a rapidly growing consumption (3). How then do these plans for increasing oil consumption relate to the supply?

An aeroplane is refuelled with primitive buckets. Oil is an energy carrier with a high energy density which can be used in most situations. It can be fairly easily utilised even in developing countries. (Photo: Per-Olle Stackman. Tiofoto).

(1) SIND = National Industrial Board

The development during the last two decades can be summarised as follows:

- during the last 20 years oil consumption in the world has almost tripled
- the annual new discoveries of oil have during the whole postwar period been between 2 and 3 thousand million tons a year
- the annual oil production has in the last few years been of the same magnitude as the new discoveries. This means that the amount of known reserves is not changing.
- at the beginning of 1978 the reserves amounted to 88 thousand million tons (17). These will last for 30 years with consumption at its present level; but only for 16 years if consumption increases as hitherto by, on average, 7.5% a year.

What does this mean for long-term oil supplies? The total amount of hydrocarbons (petroleum) available on earth is very large. Most of this amount (resource)* is not extractable, however, or else not directly accessible. The amount which can fairly easily be extracted consists of crude oil. Estimations suggest that about another 225 thousand million tons remain to be discovered and extracted. That represents about 60 years' consumption at the present level but only 27 years if consumption continues to increase as before.

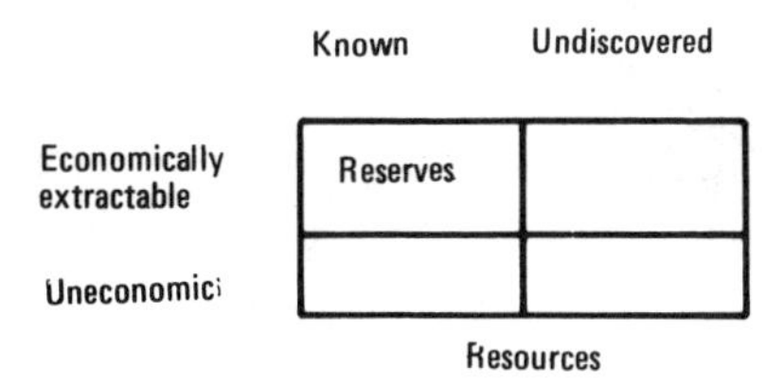

For several reasons, however, it is unlikely that oil production can continue to increase as fast as it has hitherto.

Annual new discoveries of oil may be expected to fall short of annual production in future, which will lead to a diminishing ratio between known reserves and annual production. The cause of the limited discoveries of new oil fields is that the "easy" and large fields have probably already been discovered. Those which now attract interest for prospecting and production are situated in more difficult environments, i.e. off-shore and other less accessible regions. The time lag between the discovery of a new field until production gets under way will increase, partly due to the inaccessibility of the deposits and environmental considerations.

New oil sources are increasingly being opened up outside the "old" oil fields. New states may want to have influence and control over them. That may limit the willingness of the oil companies to invest, while at the same time it is precisely the oil companies which have the resources to quickly carry out prospecting and

(1) By reserves are meant deposits in the earth's crust which can be exploited with known technology at today's prices. Apart from these reserves there are also as yet unidentified deposits. The reserves constitute a part of the resources of, for example, oil. By resources is meant the total deposits, including that which is not accessible with known technology or at competitive prices but which is judged to be potentially accessible with new technology and higher prices; but also as yet unidentified such deposits. With new technology and higher prices a part of the resources is thus converted into reserves. It should be stressed that knowledge of how the amount of reserves increases as the price rises is very inadequate (16).

production. For each oil field there are geological reasons why the ratio between reserves and production must not fall below a given value (16). The increase in demand and the size of discoveries of new reserves decide when stagnation will occur in oil production. See Figure 13.

It is apparent that only a low rate of increase in consumption and a high rate of discovery can postpone the stagnation in oil production beyond 1995. But stagnation may very well occur as early as the end of the 1980's or the beginning of the 1990's. The reason for this is not that the oil in the ground will "run out" but that the co-ordination of all the necessary activities to maintain and expand the very large production simply becomes too difficult to manage.

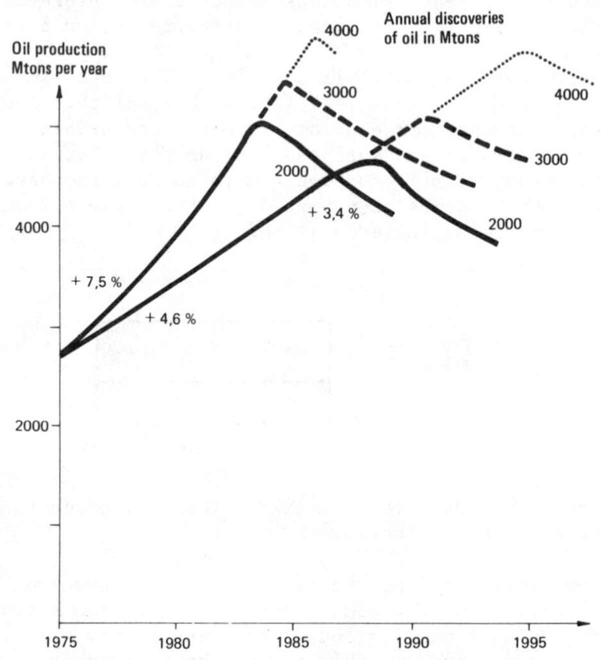

Figure 13. The development of global oil production if the ratio between reserves and annual production cannot fall below 15. Two estimates of future demand have been used. The higher one relates to a continuation of the historical growth 1950-1970. The lower one represents a projection of the plans of the OECD, the state-trading countries and a number of developing countries (16).

To sum up, therefore, we must conclude that there is a clear risk that oil production will reach its highest level in around 10-15 years' time and will then decline or level off. This would be due to the inability of the oil industry to sustain a growing oil production in the increasingly difficult environments which will then obtain. The onset of stagnation may come earlier for political reasons - if the OPEC countries introduce production ceilings (15, 18). Stagnation may begin later if the demand for oil increases less rapidly because of rising oil prices and/or a continuing or renewed international depression. The depression could also be caused (or aggravated) precisely by rising oil prices. A development during the 80's with relatively rapid economic growth <u>and</u> constant oil prices does not seem very likely.

There are considerable deposits of oil which are difficult to develop such as oil shale, tar-sand, possibly oil on the outer continental shelves, and possibilities of more effective extraction from existing fields (so called secondary and tertiary extraction) (16). There are, however, many technological, ecological and economic problems which have to be solved before oil from these deposits can play a role. An estimate by Exxon suggests that by 1990 about 1% of the world's energy requirement will come from such supplies (19).

The above argument is based on the known and probable conditions today. One can envisage (unlikely) events changing the time scale, e.g. the discovery of a new Middle East... Policies cannot, however, be based on the occurrence of such fortunate but rather unlikely eventualities.

We shall return to a discussion of the consequences for Sweden at the end of this chapter and in chapter 6.

Natural gas

Gas supplies 18% of the world's energy needs. It is easy to handle and distribute through pipelines over short distances and causes relatively limited environmental disturbances. Gas is also an excellent raw material for the chemical industry. Natural gas can replace oil, e.g. in industry, power stations, and for heating. But natural gas demands large investments for transport and distribution. There is thus a "ticket of admission" in the form of large capital outlays which every country must make in order to use natural gas. Countries with (or in proximity to) large gas deposits have so far made these investments, e.g. the United States, Canada, the Soviet Union, Britain, the Netherlands, West Germany and France. The countries in the Middle East, on the other hand, despite the fact that large quantities of gas are at present burned during oil production, have not had the industrial basis and the demand required to make a distribution system profitable.

Natural gas often occurs in association with crude oil in the oil fields. About 40% of the reserves are in fields with both gas and oil. In Iran, Alaska and northern Canada there are large deposits, but a long way away from the more important areas of consumption. The Soviet Union also has considerable supplies, about one third of the global reserve. But the deposits are located chiefly in Siberia where productive equipment is lacking and where the climate is severe. The region is expected to get its first railway in 4-5 years and there has been discussion of pipelines to Moscow and the Japan Sea.

An appraisal of the importance of natural gas as a future energy source must contain a considerably larger element of uncertainty than in the case of oil. This applies to estimates both of the available quantities of natural gas and the profitability of gas transport systems.

The gas that is used today represents about 1,100 Mtons of oil (Mtoe). The known reserves represent about 45,000 Mtons of oil and should therefore suffice for about 40 years consumption at the present level. Estimates of what may be discovered in future vary widely, but the undiscovered deposits are thought to be twice as large as the present reserves. If the oil price increases, natural gas deposits which are difficult of access will of course become profitable.

In a longer time-perspective it is possible that there are very large gas supplies in areas difficult to reach at great depths in so-called "geopressurized zones" (16). If and when these deposits can play any role is very uncertain today.

Large Swedish gas imports during the 80's must be based on so-called LNG[1] from the OPEC countries unless Norwegian fields north of 62° latitude turn out to contain very large quantities of gas. Alternatively the import could take place in the form of methanol, which would then be manufactured from natural gas in the producing country. Such a fuel could, for example, replace petrol and thus be of great significance for the endeavours to reduce air pollution in cities.

One may say in conclusion about natural gas that it will continue for a number of decades to be an important part of the energy supply of the industrialised countries. Like oil, one can expect natural gas to reach a peak of production and then diminish, first for reasons of production technology or politics, subsequently because of purely physical limitations (20). In the case of Sweden it is doubtful whether we should enter the gas market on any large scale and pay for the extensive basic investment which is needed. It is uncertain whether we can count on gas deliveries for long enough to give the gas network time to pay for itself. A Swedish LNG project would not lessen our dependence on the OPEC countries.

Coal

Let us now proceed to coal and discuss its future potential.

The developed countries were industrialised with the aid of coal and countries with large coal supplies still use coal to a large extent even if its relative share has dropped since the beginning of the twentieth century. Today coal represents 32% of the world's energy provision.

In contrast to the situation with oil and natural gas there is plenty of coal. Table 9 shows the known coal deposits in a number of regions (21).

The United States, China, the Soviet Union and Eastern Europe have nearly 80% of the world's known coal reserves.

Many predict a "coal renaissance" as coal is the only fuel which can replace oil and natural gas for a long time to come with the present technology. The supply is not an essential barrier to this. But we must then look at other aspects which could limit a boom in coal.

Coal contains sulphur, heavy metals and other pollutants which may produce serious effects on the environment. In the case of large-scale reliance on coal, strong measures are required to limit pollution, e.g. by desulphurisation of flue gases, electrostatic filters, fluidised bed combustion. New techniques must be developed. Increasingly strong environmental demands may succeed in restricting the use of coal and also push up the price level of environmentally cleaner qualities of coal.

The most serious environmental problem of coal may be the emission of carbon dioxide. Carbon dioxide is formed in all combustion of carbonaceous matter

(1) LNG (= Liquefied Natural Gas) means that the gas is cooled down (- 165° C) to a liquid, carried in special refrigerated ships and then gasified in the receiving country. Large LNG projects are being planned by Japan and the United States but the capital cost is very high. The dependence between producer and consumer is thought to be great and the flexibility low.

Table 9. Coal production and coal reserves (21).

	Coal production in 1972, Mtons	Coal reserves around 1970, Mtons	% of world reserves	The share of coal in energy consumption in 1972 %	The duration of reserves in years at constant production	Annual growth in coal production 1962-1972
United States	600	185,000	38	20	318	3.1
Western Europe	370	52,000	10	26	139	-3.8
Soviet region	520	93,000	18	39	175	2.1
China region	440	78,000	15	88	174	4.8
Eastern region	370	41,000	8	71	110	2.1
Japanese region	44	2,000	0	24	39	-4.3
Others	260	52,000	10	22	204	?
Total	2,600	500,000	100	32	194	1.6

(among others oil, natural gas, wood).[1] About half of this carbon dioxide remains in the atmosphere for a long time. An increased carbon dioxide concentration in the atmosphere will probably lead to a higher temperature, which in turn may lead to climatic changes with consequences which are, to say the least, difficult to predict. A continuing rapid increase in the combustion of oil, natural gas and coal during several more decades will probably cause serious changes in the climatic zones and therefore, e.g. in the conditions of agriculture (22).

Coal production takes place - and has done so more often in the past - in very difficult working conditions. Many labour conflicts have been concerned with conditions in coal production, but despite that not all the problems have been solved even in the advanced industrial countries (23).

Large parts of the world's coal supplies exist in areas which at present lack an industrial basis (Siberia, the Middle of the United States, parts of Australia, etc.) A major increase in coal extraction will therefore not only require investments in new mines and extensive transport systems, but new cities may also have to be built, old agricultural land abandoned, and so on. The planning periods are long, up to 10-15 years from the time the decisions are made until production in the new mines can begin (24).

The organization of the coal sector can become an obstacle to an extensive world trade in coal. Close co-ordination of investments in mines, transport and user equipment is required in order to bring about a rapid expansion of coal. In the international coal trade there are today no equivalents of the oil companies with their facilities for co-ordinating activities in different countries. Presumably the only bodies that could rapidly change this situation are precisely the oil companies. They are in fact buying their way into the coal industry in many countries; thus 14 of the 20 largest owners of coal reserves in the United States are oil or gas companies (25).

(1) It should be noted that energy production from bio systems (plants, wood), which also produce carbon dioxide during combustion, does not contribute to the increase of the carbon dioxide content in the atmosphere as an equivalent amount of carbon dioxide is consumed during photosynthesis.

The WEC Working Group on Coal has also concluded that coal will not provide a quantity of energy equivalent to present-day oil production until the mid-1980's. Figure 14 shows an estimate of future coal production and coal exports. Note the small share which is expected to be exported.

It seems most likely that the large coal producing countries today will in the first place choose to satisfy their own requirements and only in the second place and on a small scale choose to export.

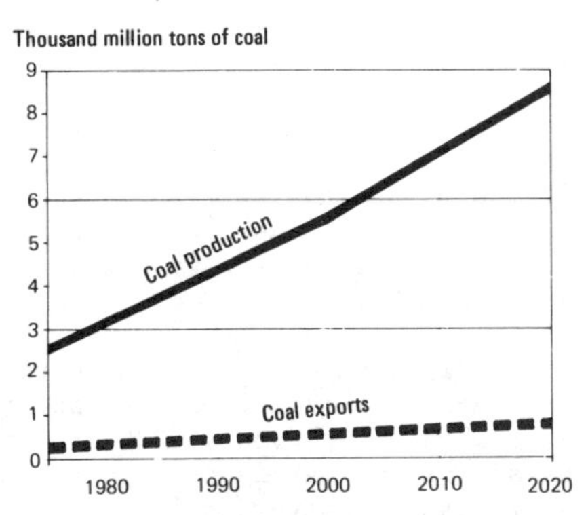

Figure 14. Forecast of the production and export of coal (24).

In the longer term there are other possibilities, but also other problems. Intensive research is being devoted to processes to produce liquified and gaseous fuels from coal, sometimes directly from coal deposits (see the section "Research, development, plans and expectations" in chapter 5). In that form coal would fit well into the distribution networks of the oil and gas companies. These processes are expensive, however, and the price of oil would probably have to rise to at least twice its present 1978 level in order for them to become profitable.

The conclusions on coal in the future are:

- coal deposits of a sufficient size are available to meet the world's energy requirements for more than a hundred years
- heavy dependence on coal as the basis of the world's energy supply involves a large risk of climatic changes but also of other environmental effects
- rapid increase of coal extraction is difficult to organize and would also cause great social and environmental problems
- the international coal market can be assumed to be of limited extent
- it is, however, quite possible that Sweden could purchase limited amounts of coal.

Nuclear power

In most countries - the United States, Britain, France, the Soviet Union and even Sweden - during the 1940's and 1950's nuclear power for energy production was closely linked to the military nuclear power projects. In the US Navy nuclear-powered ships began to be considered already before the detonation of the atom bomb in 1945 (26). The civilian reactors were developed from military prototypes. The reactor in the United States nuclear submarine Nautilus, for instance, was a predecessor of the now predominant light water reactor.

Throughout the 1940's and 1950's a large number of reactor types were studied.

Some were based on natural uranium and therefore needed no enrichment (see below) some required lightly enriched uranium, others highly enriched uranium or plutonium, others again used thorium, and so on.

Apart from the technological differences - which are relevant, e.g. to safety, the risk of proliferation of nuclear arms, and so on - there were significant differences between the roles which the reactor types could play in energy supply. Some were planned only for electricity production, otheres for electricity combined with low-temperature heat for domestic heating or for high-temperature heat for the processing industry.

When we talk about nuclear power now, 20 years later, we are essentially referring to thermal reactors of the light water type. It does not seem likely that the world could rapidly switch over to another reactor even if other types would make it easier, for instance, to deal with the problems of nuclear arms proliferation. A series of other components of the nuclear fuel cycle have been added to the reactor technology (see Figure 15).

Light water reactors use one isotope of natural uranium, uranium 235, which must be enriched before it can be used. This isotope constitutes 0.7% of natural uranium and enrichment means that the uranium 235 content is raised to 2-3%. The rest of the uranium consists of another isotope, uranium 238. When uranium 238 is irradiated with neutrons it can be converted into plutonium which, like uranium 235, can be used as reactor fuel. The plutonium can be used in all reactors but is primarily intended for use in breeder reactors. In the breeder reactor more plutonium is formed by irradiation of uranium 238 than is consumed as fuel in the reactor. In that way it becomes possible to utilise the natural uranium about fifty times more efficiently.

The nuclear power industry and the research scientists have all along seem the light water reactors as a first step in the construction of a long-term nuclear power system based on breeder reactors (28, 29).

Many studies have been made of the future of nuclear power, the latest commissioned by the World Energy Conference (WEC) (see table 10).

Table 10. Forecast of nuclear power expansion in the world in GWe. (One reactor represents about 1 GWe, so that the figures by and large correspond to the number of reactors) (30).

	1975	1985	2000	2020
OECD	68	247	955	2,423
States with central planning	7	33	402	1,610
Others	1	23	186	1,000
Total (capacity in GW electricity)	76	303	1,543	5,033

The uranium requirement for the expansion of nuclear power depends, among other things, on the extent to which breeder reactors are introduced. The WEC has assumed that the breeder reactor will be put into commercial operation in 1987 in Western Europe and in 1993 in the United States.

Despite the introduction of breeder reactors an expansion of nuclear power in line with the forecast of the WEC would involve a 15-fold increase in uranium extraction in less than 45 years. This corresponds to the discovery and location of

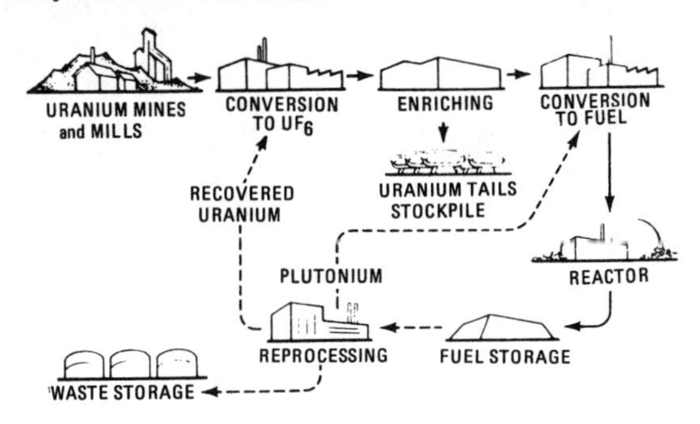

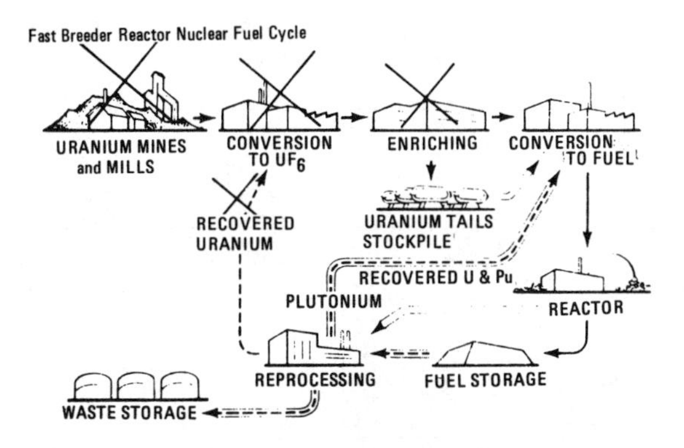

Figure 15. The fuel cycle of nuclear power (27).

more than 300 mines before the year 2015 and would involve opening up of a new mine every six weeks for the next 40 years. Without reprocessing, the uranium requirement will be higher and a new mine will be needed every third or every fourth week (30).

The reserves known at present (see Table 11) would be used up before the year 2000 if nuclear power is expanded according to the calucations of the WEC. But there are probably large unknown supplies and the search for uranium is intensifying. Thus the forecasts do not say that uranium is "running out" - on the other hand they indicate that the uranium industry may have difficulty in supplying the nuclear power stations with uranium in sufficient quantities.

To moderate such a rapidly rising demand for uranium as forecast by the WEC assumes a number of measures to ease the pressure on the uranium industry. With breeder reactors the uranium requirement will be moderated in the long run - but by how much will depend, among other factors, on how quickly plutonium is produced in the reactors. The first large breeder reactor, which is now under construction in France, Super Phenix, will produce sufficient plutonium for one additional reactor in about 30 years. That is far too long for the uranium demand to really diminish. It is through that a reduction of this so-called doubling period can be achieved only with considerable efforts and it is unlikely that it can be reduced by half before the year 1995 (31).

Table 11. Uranium supplies known and estimated at present in price classes (thousand tons) (51).

Countries	Known supplies $80/kg uranium	Known supplies $80-130/kg uranium	Estimated supplies $80/kg uranium	Estimated supplies $80-130/kg uranium
Algeria	28	0	50	0
Argentina	18	24	0	0
Australia	289	7	44	5
Brazil	18	0	8	0
Canada	167	15	392	264
France	37	15	24	20
Gabon	20	0	5	5
India	30	0	24	0
Nigeria	160	0	53	0
South Africa	306	42	34	38
Sweden	1	300	3	0
United States	523	120	838	215
	1,597	522	1,475	547
Other countries	53	18	35	43
Total	1,650	540	1,510	590

At the moment there are only three breeder reactors in experimental plants; in 1982 it is planned to start a Super-Phenix vid Malville at Isere, South-eastern France. This will be the first breeder reactor for the commerical production of energy. The picture shows the building site in summer 1977.

Among other measures which affect the need for uranium are:

More efficient enrichment which leaves smaller quantities of uranium 235 in the impoverished uranium (cf. Figure 15).

Reprocessing of used nuclear fuel to extract the remaining uranium 235 and any plutonium that has been formed.

Reactors using uranium more efficiently, perhaps in combination with thorium.

The forecast for the exapnsion of nuclear power in Table 10 may seem high but it involves a substantial de-escalation compared to before. Table 12 shows how forecasts of the expansion of nuclear power have evolved since the end of the 60's.

Table 12. Forecasts of nuclear power capacity (in GW electricity) published in various years (30, 32). (States with central planning not included).

Capacity	Time of estimate					
Year	1969	1970	1973	1975	1976	1977 (WEC)
1970	25.6	18	14	-	-	
1975	101-125	118	94	71	-	69
1980	235-330	300	264	179-192	-	-
1985	-	610	567	475-525	440-450	270
1990	-	-	1070	875-1000	750-850	
2000	-	-	-	2000-2500	1700-2000	1141

We can thus see very clearly how the forecasts have gradually been adjusted downwards.

There are many reasons for this and it is of interest to distinguish at least four of the main ones:

Technical problems. Although light water reactors have been operating for a long time the classical problems connected with them have not yet been fully solved. These include the storage of highly radioactive waste and the possibilities of achieving a good working environment in the reprocessing plants. They also include security against major disasters: a certain consensus is now emerging about the consequences of such disasters but not about the probability of their occurrence. Certain consequences, such as the evacuation of large areas for a long period, have recently attracted increasing attention. The transition from light water reactors to breeder techniques will mean that nuclear power is confronted by a further number of problems which demand technical solutions, partly having the nature of fundamental research (38, 42).

Economic problems related to the increased cost of nuclear power and the forecasts of a decrease in therate of electricity consumption. In Sweden the electricity forecast for 1985, for example, has been adjusted downwards from 176 TWh in 1972 (33) to about 120-130 TWh in 1977 (34). This reduction has naturally reduced the pressure for the expansion of nuclear power. In the United States, too, the electricity forecasts have fallen as a result of higher electricity prices and a lower rate of economic growth.

Political problems due to the fact that nuclear power technology has met with strong opposition in many countries. The effect this has on investment decisions and time-scales depends in turn on how the legal and political system functions (35). Who can appeal against siting decisions? What aspects are relevant in granting permits, and so on?

Co-ordination problems, which are largely a result of the uncertain future. The uncertainty of electricity demand creates uncertainty about the size of the required reactor industry, the capacity needed for enrichment, reprocessing, uranium mines, and uranium prospecting.

We shall deal with the last problem in somewhat greater detail as it has not been considered to the same extent in other contexts. But we wish to emphasise that all four problems are interrelated. The problems of co-ordination connected with the expansion of nuclear power are naturally aggravated by economic uncertainty and political disunity, e.g. on questions of safety during routine operations, working conditions in reprocessing plants, and the principles for final storage of waste. We shall discuss the risk of nuclear arms proliferation later, particularly in relation to different technologies for civilian nuclear power.

The reduction of future demand has led to a situation where the reactor industry now has a very large surplus capacity. A rough calculation shows that the industry today utilises about a third of its capacity (36). The uncertainty about the future role of nuclear power also spreads to the other components of the nuclear fuel cycle. Measures taken at one stage alter the conditions for other components. Slow prospecting for uranium strongly increases the pressure to establish more efficient enrichment and reprocessing, and vice versa. A ban on reprocessing will increase the pressure on the uranium industry.

If the future expansion of nuclear power is uncertain, so is that of the uranium industry. This also makes the need for increased enrichment capacity uncertain. And even if the future of nuclear power were fairly certain, the uncertainty about reprocessing creates uncertainty about both enrichment and the uranium industry. At the same time all this involves very large amounts of capital.

Here the institutional structure of the nuclear fuel cycle can be a clearly restrictive factor. Prospecting for and extraction of uranium are handled chiefly by a small number of large mining companies, of which the British Rio Tinto Zinc is the largest. The oil companies have also very quickly acquired large shareholdings - 13 of the 20 largest United States uranium proprietors are oil companies (among them Gulf, Exxon, Phillips, Atlantic Richfield). A uranium cartel, with very active involvement of certain governments, has been formed in order to stabilise the uranium market and guarantee the economy of the mining enterprises (and thereby the revenues of those countries) (37).

The enrichment stage is today controlled by national governments (the United States, Britain, France, and a joint European enterprise, Eurodif). Canada has decided in principle to build its own enrichment plant in order to export chiefly enriched instead of natural uranium.

The reprocessing stage is today controlled by governments (Britain, France, the United States) even if certain private enterprises have been involved in the United States. There are strong reasons for questioning whether the relations are not too complex to be handled by the conglomerate of private enterprises, official bodies and governments involved. The countries which have gathered all the functions under one hat, e.g. France, appear to have succeeded best. Presumably the organization of the electricity sector has also been of some importance. Britain, France and Sweden, which have large-scale nuclear power programmes compared to their populations, all have a heavily concentrated or nationalised electricity sector. That has made it easier to organise co-ordination of development and purchasing than in, e.g., the United States (35).

The importance of the nuclear power programme for the proliferation of nuclear arms is another element in the political picture. The possibility of developing nuclear warheads for governments and sub-national groups exists in several ways.

One means of producing nuclear warheads is by enrichment, and several new methods are being developed which simplify this (centrifuge, jet nozzles and laser methods) (43). Another means is by way of reprocessing; small reprocessing plants merely for arms production have become a possibility (44). The number of means is growing with the expansion of civilian nuclear power. If the latter had not existed, the inducement to develop new technologies, e.g. for enrichment, would have been considerably less.

The use of breeder reactors involves very extensive handling of plutonium and thus poses very great security demands so that this plutonium does not get into the wrong hands.

The difficulty in removing the military material from civilian nuclear power programmes depends partly on how it is designed. The combination light water reactors-breeder reactors (LMFBR)[1] is hardly the one involving least problems in this respect, as the plutonium stream occurs to such a great extent outside the reactor itself.

In actual fact the problem of minimising the risk of proliferation of nuclear arms has never exercised the nuclear power industry. Alvin Weinberg has written:

"We have never seriously considered resistance to proliferation as a design criteria for reactors. Had it been a design criteria from the beginning I suspect our lines of development may have been very different - for example, my favorite, the molten salt reactor, might have been taken much more seriously". (28)

The international future of nuclear power is thus not free from problems. A number (though not all) of the difficulties derive from the nuclear power industry's own ambitions to rapidly develop techniques for plutonium handling (reprocessing plant etc) in order subsequently to move on to a breeder economy on a large scale.

If the idea of breeder reactors on a large scale is abandoned - or at least if the decision is postponed - the political possibilities for nuclear power will presumably be widened in many countries. There will then be several technical means of increasing the yield of energy from a given quantity of uranium by as much as 2-4 times. One could also use thorium as a fuel. The development of these reactor types on an industrial scale takes time, however.

Development may follow one of the following lines:

A development from light water reactors towards breeder reactors in line with the WEC's study and the plans of the nuclear power industry. Very strong commercial interests are pulling in that direction, but for them to succeed the political problems and the co-ordination problems of the nuclear power industry must be overcome.

A ban on reprocessing, while thermal reactors are accepted. Reactors using uranium more efficiently can gradually be introduced, may be also based on thorium. Nuclear power could then play a role in the world's energy supply, but only a limited one for a fairly long time. The pressure on uranium extraction will be heavy.

(1) LMFBR stands for Liquid Metal Fast Breeder Reactor, i.e. reactors in which liquid metal, e.g. sodium, is used as cooling agent.

Nuclear power capsizes wholly or partly. The international recession has caused a reduction in the rate of increase of electricity demand. Nuclear power as a link in the energy supply is a uniquely rigid system. The capital intensity, planning periods and the size of the individual components make the nuclear power industry extremely badly equipped to adapt to uncertain markets. There is a very definite possibility that the nuclear power industry in some countries may collapse.

Renewable energy sources

The potential of renewable energy sources is broadly illustrated in Figure 1 (p.5), from which will be seen, among other things, that eight times the amount of energy which is sold in the energy markets is absorbed in processes of photosynthesis. The amount of renewable energy does not constitute a limitation. The difficultires are rather technical and economic. The energy flow is diluted, which means that large areas must be used.

It is only since the oil crisis of 1973-74 that development work with a more serious purpose has begun around new techniques to utilise renewable energy sources. Despite the short time a large number of promising results and ideas have emerged.

We shall here restrict our attention to a short description of a few technical proposals. Several of the techniques mentioned below are not at present in commercial production. But it has been shown that they work. The uncertainties are mainly economic but there are also environmental effects.

Hydro power has long been used and there is a large potentially above all in developing countries. There are, however, definite ecological problems, mainly due to silting.

Wind power is a possibility. Studies indicate (46) that both very large (1-4 MW) and small (10-50 kW) systems could be economical.

Local solar heating (and cooling) is being studied in many countries and may be economical for a large part of the earth's population. The major problems derive from the fact that present buildings are not adapted to this technique.

Electricity generation from solar radiation has been discussed in various technical forms. Photoelectric cells (solar cells) are being studied, for instance, in which the solar radiation is directly converted by means of semiconductor techniques into low voltage direct current. It seems that the costs are very quickly being reduced. Experiments are also being undertaken with solar towers, in which the solar radiation is focussed by mirrors into a chamber where steam is generated. The steam is then made to drive a turbine. The solar towers are mainly conceivable in very sunny regions (the south-west of the United States, the Sahara, certain parts of Spain, south-eastern Europe etc.)

Biomass (i.e. plants and animals) is already utilised and new systems for industrial application are being studied in several countries. Brazil already has a very extensive programme for the production of ethanol through fermentation of, among other plants, sugar cane. In the United States and Sweden studies and research work are being conducted. Cultivation both on land (fast-growing deciduous trees) and in the sea (kelp, algae) has been discussed. The essential conditions differ markedly from one country to another, not least because of the competition for arable land.

The possibilities of utilising nature's energy flows and different techniques for the purpose vary from one country to another. Certain developing countries - e. g. India - even today use about 70% non-commercial energy (wood, dung etc) and their immediate problem is rather to avoid negative effects from it, e.g. in connection with over-cutting of trees and bushes.

But the position of industrialised countries also varies considerably. In Table 13 a few countries are compared.

Table 13. Intensity of energy consumption related to solar radiation for different countries (1973). Land area requirements (in % of total area) to cover energy needs (1973) through renewable energy techniques at different levels of efficiency of energy extraction (1% and 10% respectively. For comparison it many be noted that solar heat installations have an efficiency of extraction amounting to 30-40%, solar cells 10-20%, and plants 0.5-2% (e.g. energy forest plantations).

	Energy use (2) kWh/m^2 and year	Solar radiation kWh/m^2 and year	Percentage of the country's land area necessary at an efficiency of extraction of	
			1%	10%
United States	2.1	1,800	12	1.2
France	3.3	1,300	25	2.5
West Germany	11.6	1,200	94	9.4
Japan	8.7	1,100	78	7.8
Sweden	0.84	1,000	8.7	0.87
Algeria	0.025	2,200	0.11	0.011
Saudi Arabia	0.038	2,400	0.016	0.0016
India	0.26	2,000	1.3	0.13
Tanzania	0.0094	2,000	0.048	0.0048

Sweden is favoured in comparison with other countries - the large land area compensates more than well for the somewhat lower solar radiation. One conclusion is thus that studies specific to certain countries must be undertaken. Today there are very few such studies.

The debate on renewable energy sources has generally started from the assumption that they must be internal and national. There are also ideas based on the conception that solar towers in the sunnier parts of the earth (for example the Sahara) would produce very large quantities of energy which is then transported in the form of high voltage electricity or perhaps as hydrogen gas (54). Solar cells can presumably be made more competitive by means of concentration techniques. It should therefore not be excluded that a world-wide trade and a new international division of labour may develop between countries which are rich and poor in solar energy on the basis of certain kinds of solar technology. Such a development is several decades away, however.

In conclusion, the utilisation of renewable energy sources on a large scale suits certain countries better than others. High solar radiation provides some advantages, large land areas (e.g. for biomass) provide others. One cannot automatically assume, therefore, that solar energy is only suitable for countries around the equator. It is rather the case that countries like Canada, the United States and Sweden have considerable more potential than, say, France and West Germany. Solar energy may also have definite advantages for many developing countries as compared with energy systems based on natural gas and nuclear power (11). Electricity from photovoltaic cells or wind power does not require the same transmission and distribution systems as electricity from nuclear power stations. Methane gas (produced from biomass) does not require an extensive gas distribution system.

Other energy sources

Geothermal energy now plays an important role in a few countries but, in order to be utilised more generally, it demands technological innovations, e.g. in deep drilling techniques. If and when these may be achieved is uncertain. Unitl then geothermal energy remains an important energy source locally but cannot cater for the main part of the world's energy supply.

Fusion has often been presented as the future solution of the world's energy problems. In terms of raw materials there are no real limitations. Several fundamental problems at the basic research level still remain to be solved, however before the principle can be reagrded as deonstrated (47). For the construction of power producing installations it is also necessary to solve other problems of a basis research nature. Thus the very intensive neutron radiation requires the development of new materials. Even if fusion power does not produce the same waste as ordinary nuclear power, very large amounts of radioactivity will be generated. Fusion is not a "clean" energy source. It is highly uncertain whether and when fusion could contributed to the energy supply.

In the case both of fusion energy and geothermal energy, even after fundamental breakthroughs have been achieved, it will take several decades before they can contribute in a significant way to global energy supplies. The long-term planning of today cannot presume that these energy sources will become available.

Sweden in the future energy market

Energy sales in Sweden are dominated by oil. Next comes hydro power. Oil, coal and nuclear fuel are imported. The coal is mainly used in the iron and steel industry. The detailed composition of the supply is hown in the following table.

Table 14. Supply of energy to Sweden 1976 (48).

	TWh	%
Oil	313	72
(of which for electricity production)	(32)	
Coal, coke	18	4
Lyes, wood, waste	35	8
Hydro power	54	12
Nuclear power as electrical energy	15	3
(as thermal energy 45 TWh)		
Net import of electricity	2	0.5
Total energy supplied	437	100

The oil in the table is not a unified concept. Crude oil constitutes 44% of imports. The remainder consists of refined products like petrol, paraffin and various grades of heating oil.

Crude oil is imported from 13 countries. The largest exporter of crude oil to our country is Saudi Arabia (2.8 million tons) followed by Iran (2.6 million tons), the United Arab Emirates (1.9 million tons), Nigeria (1.8 million tons) and Qatar (barely 1.5 million tons). Refined oil products are imported from about 15 countries. The major source in 1976 was the Soviet Union, 3.9 millions m^3, and Britain, 3.1 millions m^3. Petrol is imported chiefly from refineries in Denmark and in varying amounts from about ten other countries (49).

Uranium is imported chiefly from Niger and Gabon through French companies. In addition uranium comes from Canada and the United States. Enrichment is carried out chiefly in the United States, while a small amount will be enriched in the Soviet Union (50).

That is, in broad outline, a view of the dependence of Swedish energy supplies on foreign sources. For the future it will therefore be important to decide both what foreign dependence we should (or must) have and how the conditions for energy supply may develop globally. We shall start with the latter.

Future world energy supplies should be seen in two time perspectives. The shorter one covers the period roughly to the end of the century. There are definite risks of disturbance to supplies in that period. In the longer time perspective stretching beyond the turn of the century oil and gas, on the present evidence, will have to be gradually replaced as the basis of the world's energy supply. It is conceivable that oil from oil shale or tar-sand, gas in geological "pockets" ("geopressurized zones") and other sources which today may seem exotic can prolong the oil and gas epoch. The technical, economic and environmental conditions have not been closely investigated and are obviously very uncertain.

In the long perspective, and for the world as a whole, there are only three relatively credible substitutes for oil and gas:

Coal exists in very large quantities and its utilisation is presumably restricted mainly by environmental disturbances.

Nuclear power could replace oil globally but that would require a transition from the thermal reactors of today to breeder reactors.

Renewable energy sources have only been developed fairly intensively since the oil crisis of 1973-74 (apart from hydro power and wood). Many applications have not yet reached technical maturity, but many promising results have been achieved.

Even these three alternatives are at present uncertain. One can neither conclusively decide in favour of or conclusively dismiss any of them. In the shorter time perspective - the period up to the end of the century - we must take into account the long-term alternatives.

As we have argued earlier in the chapter (and in a separate report) (16) there is a great risk that oil and gas production will stagnate during this period, in the case of oil perhaps even before 1990. If the oil-consuming states do not plan in advance for such a development, the situation can easily become threatening. The competition for oil will then become fierce, not only between industrialised and developing countries but also between the industrialised countries. This might well influence also their mutual political relations. If the United States, Japan and Western Europe are competing for the Middle East's oil it will become a weighty factor in their foreign policy. There is also a real risk that the United States, Japan and the powerful states in the EEC will then primarily ensure their own supplies, while small and weaker countries are left in the lurch. The result for Sweden could be both less oil and more expensive oil. One should not assume in the case of Sweden that Norwegian oil will significantly ease the situation. An unplanned stagnation of oil consumption will increase the demands on Norway to extract oil rapidly. Sweden will then have to compete with all other countries.

But rising oil prices also put the economies of the industrialised countries under severe stress. The threat of balance of payments problems may lead several countries to adopt an economic policy directed towards strengthening exports and reducing imports. This in turn could well lead to a lengthy depression. Sweden could be hard hit even if we ourselves conduct a successful policy of phasing out oil. Our dependence on oil therefore derives not only from its importation but also from our general economic dependence: our integration with the economies of other countries, which are in their turn dependent on oil imports.

An induced stagnation can thus become a threat both to the world economy and to political stability. But stagnation can also come about entirely calmly and undramatically provided that the main oil-consuming countries consciously plan for that contingency. They can do so by replacing oil with other energy sources, by conservation, and by consciously aiming at lower oil consumption in their economic development. A planned stagnation is no threat, an unplanned one is.

One cannot help wondering whether the energy supply of Western Europe during the 80's and 90's does not contain other patterns of conflict. Three small states - Norway, the Netherlands and Sweden - have a surplus of oil, gas and uranium respectively. Two large states - West Germany and Britain - have a surplus of coal. The other European countries lack energy reserves of major importance. The differences between the countries with a surplus and those with a deficit happens to coincide with a number of other differences (political, religious, curltural etc) and could strengthen a tendency to division of Europe on north-south lines, with one grouping around the North Sea/Baltic and one around the Mediterranean (55).

We can now ask oursleves how other forms of energy will be affected if oil production is expected to stagnate. Interest will naturally increase in the fuels which in this time-perspective can complement oil: coal, natural gas and uranium. In the United States the Carter administration has tried to reduce its oil imports by replacing oil and natural gas with coal in large industrial enterprises and power plants, while simultaneously in part replacing coal with nuclear power.

A somewhat similar development is taking place in other OECD countries. It should be noted that countries which have large supplies of coal use oil preferably where it is difficult to replace - in the transport sector and in the form of light fuel oils - while the large consumers use coal. It is presumably, in particular, the countries with little coal which are best able to replace oil by coal. Countries like the United States and West Germany would thus find it more difficult to replace oil by coal than, for instance, Sweden and Japan.

The output of uranium (for nuclear fuel) seems destined to be a bottleneck (see p. 52). In that case the nuclear power investments of various countries will both have an influence on each other and at the same time alter the market for coal, oil and gas. President Carter has expressed intentions of letting light water reactors expand in the United States. If he succeeds, pressure will be created on the uranium market. If he does not succeed, the expansion of nuclear power in the United States will be slowed down. Then two things will presumably happen. Firstly the demand for coal will increase and it appears even more unlikely that the United States would start exporting coal. Secondly the United States nuclear power industry would need to be gradually phased out. This would in itself reduce the pressure on the uranium market. But the consequences for Western Europe of both the absence of United States coal exports and the collapse of the nuclear power industry would appear to be, to put it mildly, difficult to foresee.

If nuclear power expands slowly in Western Europe much oil, coal and natural gas will be required, above all from the EEC countries. As internal coal extraction within the EEC involves problems with the environment and labour, the demand will presumably be directed mainly towards the international coal trade, in the first place to Poland, South Africa and Australia. Sweden's opportunities for importing coal will be reduced if the EEC countries begin to appear on the coal market.

The outlook for natural gas is broadly parallel. There are two large surplus regions in the World, the Middle East and the Soviet Union. The natural gas market, however, is tied to the pipeline system (or the LNG systems). As distribution requires large investments it is likely that long contracts will be sought. The uncertainty for the EEC countries will be politically conditioned: to increase the dependence on the Middle East or exchange it for dependence on the Soviet Union?

If, on the other hand, nuclear power is accepted on a large scale in Western Europe, the pressure will ease on coal and natural gas. Sweden's opportunities to complement oil with coal or natural gas at reasonable prices would then increase. But at the same time it is to be expected that the supply of uranium will become a limiting factor. Thereby Swedish uranium policy could gain significance in an international context. We shall present certain speculations on this matter.

The uranium trade must primarily be seen in the context of the proliferation of nuclear arms. Widespread investments in reprocessing and breeder reactors would make the problem of preventing the spread of fissile material, particularly plutonium, much more difficult even than it is at present. Investment in the breeder reactor is, however, under serious discussion in, for example, West Germany, Britain, France and Japan in view of their lack of uranium supplies of their own. It may be discussed whether these countries could postpone decisions to develop the plutonium cycle (see p.52) or at least refrain from spreading the technology to other countries if their uranium supplies were guaranteed through some form of international agreement between importers and exporters of uranium. One condition would be a co-ordinated export policy between uranium producers - a "UPEC" - consisting primarily of Canada, Australia, the United States, Namibia and, possibly, Sweden. The political conflict around nuclear power in general and the breeder programmes in particular is so fierce in countries like West Germany and France that an agreement which guaranteed the supply of uranium in exchange for a postponement of the plutonium cycle might in fact have its attractions.

There are naturally only hypotheses. Even greater is the uncertainty about the role that Swedish uranium could play. The Ranstad deposit represents 15% of the world's known uranium reserves and about 75% of Europe's. The presently known domestic uranium reserves would be sufficient to produce around 400 TWh of electricity (about the whole of Sweden's energy consumption) for 30 years in light water reactors. This requires uranium extraction eight times greater than in the Ranstad project which is under discussion. But we can turn the question around. Can we act in so narrowly national a manner even if we should wish to? How large international significance has the Swedish uranium? Does the Ranstad deposit provide opportunities not only to supply a Swedish reactor programme with uranium but also to export uranium? Swedish imports of oil, coal and natural gas, and the export of uranium, should in that case be analysed in conjunction. Bilateral agreements with countries which require uranium but have their own deposits of oil, coal or gas might then be attractive. In order that the Swedish uranium resources may have any importance in such a context or in relation to a "UPEC", however, it will be necessary to plan the uranium mining at Ranstad on a considerably larger scale than has hitherto been discussed. Mining on the scale of five "Ranstad projects" (5 x 1,300 tons) at the beginning of the 90's would then represent at most 5% of the projected global uranium requirements (51). Such a policy would obviously limit the possibilities of a large Swedish nuclear power programme (compare chapter 4).

An additional reason for looking at trade in oil, coal and natural gas from an overall perspective is the rapidly growing interest of the international oil companies in coal and uranium (25). The oil companies already now control the distribution system for that part of coal production which may come to consist of gas or liquids. They are also reinforcing their influence over direct coal production (25). Such a development may be the only feasible one if it is to be possible for coal to replace oil to any large extent: none but the oil companies have the experience of both chemistry and distribution at the international level and on the large scale which is required for such a task.

We could thus become strongly dependent on the administrative capability of the oil companies and on their ability to exchange different fuels for each other. This may also be of importance for energy prices. That the oil corporations collaborate in questions of price fixing is quite obvious - OPEC was not the first oil cartel (52, 53). The cartel on the uranium market, which was discussed in the section on "Nuclear power", has several oil companies among its members.

There is thus a clear evolutionary tendency for the oil companies to preserve their role as the principal link in the international energy trade and, in addition to extend further into individual countries and to new markets such as coal and uranium. As opposite parties they then have the governments of the industrialised countries, which are to some extent stronger than the governments of developing countries which they previously had to deal with. But there is undoubtedly a need for a strong coalition of governments of the industrialised countries if the oil companies are not to dominate the stage.

We have not taken into account here the renewable energy sources during the 1980's. The reason is not that we regard their development as technically unrealistic but, quite simply, that at present there are not powerful forces pulling in that direction in the countries which have the greatest importance for conditions on the world market. A change may take place, but we cannot presume that it will occur.

Our international survey has underlined the thesis that oil dependency is the dominant problem. It should be seen in two time-perspectives. In a shorter perspective (10-15 years) it is necessary to reduce the pressure on the international energy market in order thereby to reduce the risk of crisis which may have serious consequences for our country among others. In a longer perspective it is necessary to allocate a subordinate role to oil in Swedish energy supplies.

The survey also shows quite clearly that Sweden must acquire an all-round view of the international market for fuel, including uranium. This must comprise the developing countries. The institutional forms, for example the development of the oil companies into energy companies, play a vital role for both the developing countries' and our own supply. For the latter we may finally refer back to the three alternatives mentioned above (p.61).

There are presumably forces which, under certain conditions, could in the long run pull us into a coal economy, but the uncertain market prospects, the dependence on foreign sources and, not least, the possible long-term effects on the climate (22) make it hardly plausible to base Swedish energy policy on coal in the long term

An extensive nuclear power system which does not presuppose a speedy transition to breeder reactors would be conceivable for Sweden during a limited period. It would require that we mine our known uranium reserves, but export very little or none at all. If there is more uranium in Sweden, this period could be extended. (It will, however, be necessary to decide at an early stage to introduce breeder reactors, as it takes time to develop installations for reprocessing, plutonium storage etc). Whether such a nuclear power development, differing from that of the world around, is realisable, is difficult to know. If the uranium demand in the world becomes large, and we are then already mining uranium, export may become unavoidable.

It also seems possible to base Swedish energy supplies solely on renewable energy sources. Such an alternative could be based on domestic sources and - by and large - domestic technology. The latter is still rather undeveloped, however. But this could provide the opportunity for Sweden to choose its own direction.

In the next chapter we shall study more closely the two long-term alternatives for Sweden - the nuclear alternative and the renewable alternative.

4 Two Energy Futures

In this chapter we shall look at how two essentially different energy production systems might be designed in a future Sweden. Two extreme alternatives are presented: Nuclear Sweden, in which energy production is chiefly based on nuclear power, and Solar Sweden, in which only renewable energy sources are used.

To be in a better position to compare the two alternatives we shall proceed from a number of assumptions about society in the year 2015. The society is of the present-day type and the economic development which has led to this society has been based on a growth rate of about 2%. No fundamental economic changes have thus taken place either in Nuclear Sweden or Solar Sweden.

The choice of this particular point of departure does not imply a stand in favour of or against changes in the contemporary society, e.g. toward a society which consumes less resources and energy. The object is rather to simplify the debate i.e. to avoid having to conduct a discussion about fundamental societal changes simultaneously with a discussion of future energy systems. Nor must points of departure be regarded as forecasts.

Both alternatives contain uncertain components which are difficult to assess with regard to technical performance, costs, health, environmental and safety effects, but we have endeavoured in all cases to use cautious assessments.

We are thus dealing with a number of uncertain factors, so that, in our opinion, it is too early at present to make a choice which precludes one of the alternatives The important thing under the circumstances to preserve one's freedom of action, which can only be achieved by an active and purposeful policy. Nuclear Sweden is probably close to the energy system that we will get if no political decisions are taken to follow another direction.

We consider it important to try to present an overall picture and our hope is that both Nuclear Sweden and Solar Sweden may provide a basis for discussion about what people wish to see as possible in the future, especially as both Nuclear Sweden and Solar Sweden can be regarded as representing different values and opinions which are current in the country today.

Point of departure

1. The time is the year 2015.
2. The population of Sweden is the same as today.
3. The number of dwellings has increased by 40%; the total dwelling space has increased somewhat more. At the same time the specific energy consumption diminshes by 30% (energy requirements per m^2 of dwelling space).
4. Total production both of commodities and services has doubled.
5. The energy consumption per unit produced - the specific energy consumption - has diminished by 20% in industry and by 50% in the service and transport sectors.

The time-perspective is set at 35 years. There are several reasons why oil consumption ought to be limited within not too many decades. A reasonable time for doing so appears to be 35 years. We have not made a forecast of what the price of oil might be and still less investigated whether the energy systems discussed will be more or less expensive than oil in 10 to 20 years. It is not possible to make reasoned assessments, and speculations may be misleading.

No economic balance has been drawn between the costs of supply energy and the costs of saving it. We have been content with a general assessment of whether the supply systems proposed by us will be feasible within the national economy of 35 years hence.

The assumption in point 4 means that production will increase by somewhat less than 2% per year, which may be compared for instance with the figure used by the 1975 Long-Term Planning Committee, which assumes an average growth of 3% up to the year 2000.

The amount of "utilised" energy is the same in both cases, i.e. it involves the same amount of heated floor space, transportation, production of steel, paper and pulp, etc. Roughly the same amount of energy is thereby produced in both alternatives, even taking into account that electricity might be used more efficiently than fuel in certain processes.

The assumption in point 5 could, for example, be compared to the fact that specific energy consumption in industry has fallen by about 30% during the last 20 years despite relatively decreasing energy prices (1). With these assumptions we have estimated the energy use in the year 2015.

Table 15. Assumed distribution of energy consumption in different sectors in the year 2015.

	1975 Energy TWh	2015 Volume of production relative to 1975	2015 Change in specfic energy consumption relative to 1975	2015 Energy TWh
Production of goods	165	+100%	-20%	264
Service production				
Communications	75	+100%	-50%	75
Other	70	+100%	-50%	70
Housing, including domestic electricity	80	+ 40%	-30%	80
Total final consumption	390			489
Conversion losses	25			60-80
Total supplied	415			550-570

The energy requirement for the year 2015 is thus 550-570 TWh, which is by and large the same as the forecast of the National Industrial Board (autumn 1977) for 1990 plus a continued annual increase of barely 0.5%.

The production volume for the year 2015, which underlies these calcuations, can be used in different ways. Everybody can enjoy twice their present living standards. With an equalisation of standards everybody could enjoy a standard more or less equal to that enjoyed today by the tenth which consumes most (39). The increase in production could also be used for a substantial increase in foreign aid.

It should be stressed that the production level and energy consumption here selected do not constitute a prediction that this will actually be the situation in Sweden in the year 2015. It is a means to concretise what different energy supply models would involve. The social structure is consciously selected as a projection of the present one. A development aimed at achieving lower levels of energy consumption would naturally make it easier to introduce a new energy system as a substitute for oil.

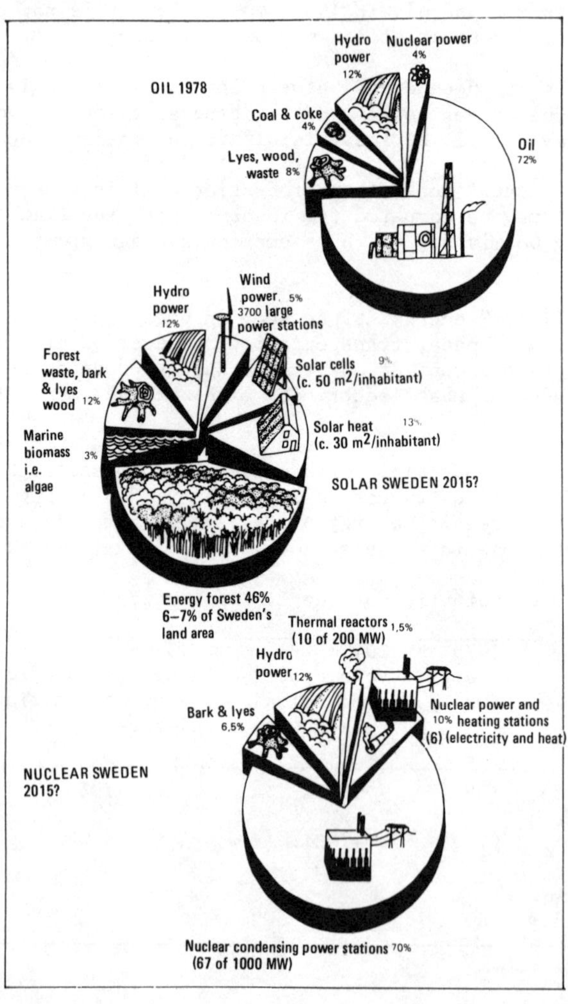

Figure 16.

Nuclear Sweden

Production Plant

For our model of Nuclear Sweden we have taken the reactor types which are furthest advanced in the plans of various countries and can thus be regarded as "least unkown technology". The energy supply of Nuclear Sweden is thus based on an increasing number of light water reactors (LWR) for the production of electricity and heat. They will later be supplemented by fast breeder reactors (FBR).

Apart from nuclear-power-based energy hydro power will contribute 65 TWh per year (according to the parliamentary decision of 1975) and biomass (bark, lyes) 36 TWh per year. Both figures correspond to the present level.

Industry, housing and services are supplied with electricity from nuclear condensing stations, with hot water and electricity from nuclear combined heat and power stations, and hot water from nuclear heating stations. District heating is delivered from 6 nuclear co-generation plants (900 MW electricity and 1000 MW heat). In addition thermal reactors (200 MW heat) are conceivable in about 10 further localities where they can meet the bulk of the heating requirement. Other energy requirements for the heating of buildings and industrial use are assumed to be covered by electricity.

This provides the supply system shown in table 16 where nuclear condensing stations must deliver the remaining quantity of energy, i.e. about 380 TWh per year. With a usage time of 5700 house (7, 8, 9) and with a block size of 1000 MW this would involve 67 reactors.

A more complete picture of the energy system is given in Figure 19.

For this programme to be realised by 2015 an extensive development of nuclear power after 1990 will be required. This is assumed to take place as in Figure 17. The first reactor in Nuclear Sweden is put into operation in 1989, which ought to ensure that the present capacity on the industrial side can be sustained.

The 13 reactors in the nuclear programme decided upon by the 1975 Parliament will have become obsolete by 2015 and must therefore be replaced by then. They therefore do not form part of Nuclear Sweden.

This nuclear power programme will make a contribution to energy supplies as shown in Figure 18. That gives some idea of the scaling down of other energy types, oil above all. We have assumed that the total energy requirement will have developed according to the forecast of the National Industrial Board (alternative A) (2) up to 1995. After that it will increase by barely 0.5% per year in accordance with the discussion in the introduction to this chapter.

Fuel

In chapter 3 we touched on the situation of the international uranium industry and our assessment was that the development of the nuclear power programme of Nuclear Sweden must be based primarily on domestic resources.

A light water reactor with a capacity of 1000 MW uses about 4600 tons of natural uranium (9) during its life span (which is assumed to be 30 years). The Billingen deposit, which is estimated to contain 300,000 tons of extractable uranium, would thus suffice for about 65 light water reactors for 30 years without reprocessing of the fuel and for about 95 reactors if reporcessing and recycling of uranium and plutonium take place (10). This is clearly not enough for a lasting and extensive Swedish nuclear power programme.

Table 16. Nuclear Sweden - total energy production in 2015.

	Energy carriers, TWh	
	Electricity	Heat
Hydro power	65	
67 nuclear condensing stations, 100 MW, average usage time 5700 hours	382	
6 nuclear power and heating stations, LWR, 900 MW electricity, 1000 MW heat 5700 hours	31	24
10 thermal reactors, 200 MW, 4000 hours' usage time		8
	478	32
Biomass, bark and lyes		36
Total amount of energy supplies	546 TWh	
Losses etc.	58 TWh	
Total amount of energy consumed, rounded off	488 TWh	

The possibilities of producing uranium in large quantities from other deposits than Billingen are very uncertain. There is uraniferous alum-shale in several parts of Sweden, including Scania (Osterlen), Oland and Gotland, in the mountain chain, and under the seabed in the Bothnian Sea. The uranium content is, however, inadequately known at present and presumably low. Figures of 50-250 ppm (millionths) have been quoted (11). The Ranstad shale has a uranium content of about 300 ppm.

The chemical process required to extract uranium from shale has not yet been tested on a full scale. The economy and environmental effects of the process are consequently also uncertain factors and one should remember that very large quantities will be required if shale with a low uranium content is to be processed (12).

Uranium is, however, utilised over 50 times more efficiently in fast breeder reactors. The uranium from the Ranstand mine at Billingen would, if used in fast breeder reactors, suffice for about 100 such reactors for 1000 years, which seems a reassuringly long time. The fast breeder reactor option is thus quite feasible from this point of view and is a natural continuation of the development if reprocessing of spent fuel takes place.

An extensive Swedish nuclear power programme must also be viewed as part of an international effort and in that context it is doubtful if Sweden can reserve its uranium deposits for putely domestic use.

There is a connection between light water reactors and fast breeder reactors. The latter require plutonium (or highly enriched uranium) to start them up. Plutonium does not occur naturally on earth. It is produced from uranium 238 in light water reactors, but also in fast breeder reactors. The characteristics of the fuel cycle in the combined LWR-FBR system simply mean that the introduction of fast breeder reactors is limited by the amount of plutonium derived from the reprocessed fuel of the light water reactors. About 25 light water reactors produce sufficient plutonium per year to start up one fast breeder reactor a year and we can calculate

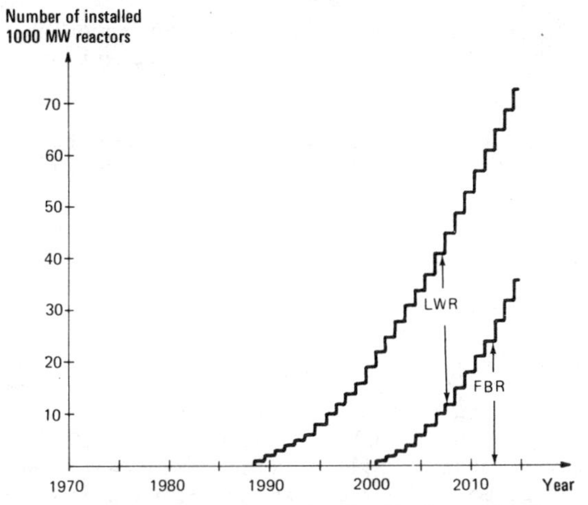

Figure 17. Rate of development for reactors in Nuclear Sweden. Distribution between light water reactors (LWR) and fast breeder reactors (FBR). Reactors existing at present or now under construction are not included.

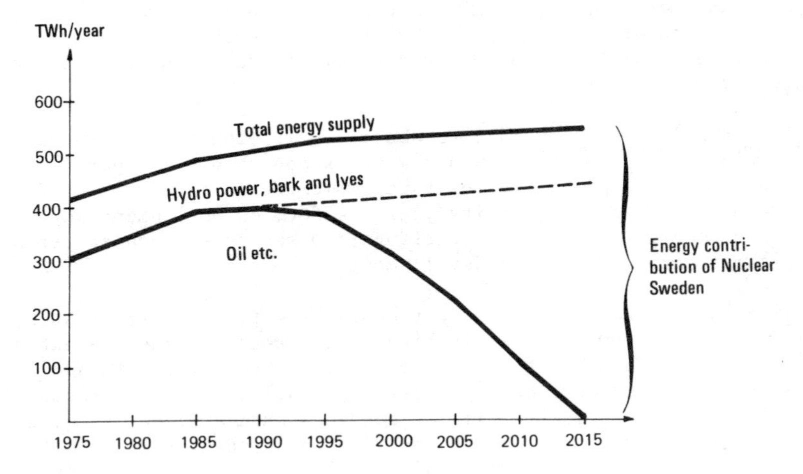

Figure 18. The contribution of Nuclear Sweden to Sweden's energy supply up to the year 2015.

exactly when we have sufficient plutonium to start up a fast breeder reactor. The nuclear power programme of 1975, which has a total capacity of about 10,000 MW, could thus during its life time produce plutonium for about 12 fast breeder reactors. For reasons of plutonium production the proportion of fast breeder reactors by 2015, therefore, cannot be greater than we have assumed. This, furthermore, presupposes that all the plutonium produced by the reactors, including that envisaged in the decision of 1975, is retained and utilised.

We have reckoned on fast breeder reactors being introduced in Sweden from the year 2000, at the rate indicated in Figure 17. By 2015 about half of the reactors will be fast breeder reactors. The proportion of fast breeder reactors will subsequently increase in step with a possible rise in energy demand or because the light water reactors have become obsolete and need to be replaced. Whether a possible increase in energy demand and replacement of obsolete reactors should be met with light water or fast breeder reactors after 2015 will depend on the uranium situation. Scarcity of uranium will probably lead to a development of Nuclear Sweden toward a Plutonium Sweden.

The nuclear power programme of Nuclear Sweden will create a uranium requirement for the light water reactors of a total of 100,000 tons up to the year 2015. The quantity of plutonium in the system will be about 100 tons by 2015. The time schedule for the fuel cycle could become very tight and this could in actual fact constitute a bottleneck.

Fuel Cycle

Uranium mining. The Supply Group of the Energy Commission believe that if prospecting for further deposits is intensfied it ought to be possible to have 2 or 3 other uranium mines working by the beginning of the 1990's apart from Ranstad and Pleutajokk. Production could reach a total of around 2000 tons of uranium/year (13). This quantity of uranium will suffice comfortably for the 1975 reactor programme. In order to meet the requirement for natural uranium in Nuclear Sweden a mine of Ranstand's size (1300 tons of uranium per year) will have to become operational around 1990. In addition 9-10 mines with a capacity of about 400 tons of uranium per year will be needed. With a uranium content of about 300 ppm (as at Ranstad) that means mining of about 16 million tons of shale a year, equivalent to half of the iron ore mining in the country (about 35 million tons a year). Added to this there will be a need for a chemical processing industry to extract the uranium from the shale.

The enrichment requirement amounts to about 3.6 million separation units. One enrichment plant ought to cover Sweden's needs and should be operational by 1990. An installation of the gas diffusion type, which would cover all of Sweden's needs, consumes about 10 TWh of electricity/year. A form of enrichment which uses considerably less energy is the gas centrifuge. Other types such as jet nozzles or laser enrichment are still under development.

Reprocessing is required of fuel both from light water and from fast breeder reactors. All the plutonium produced in light water reactors must be extracted and stored in order subsequently to start up the fast breeder reactors. By 2015 about 1500 tons of spent fuel must be reprocessed annually, including about 400 tons of breeder fuel. One installation will be needed around 1997 for the light water reactor fuel (110 tons/year), one around 2006 for breeder fuel (400 tons/year) and after that yet another around 2013 (which will then replace the oldest one).

The risk of plutonium going astray has led to the suggestion that plutonium handling should be concentrated geographically and that transport should be avoided as far as possible. The reprocessing plants should therefore be located together with the fast breeder reactors. For the 36 fast breeder reactors operating in 2015 more than one reprocessing plant would then be required. A possible solution is to locate together about 6 fast breeder reactors with a (small) reprocessing plant for 60 tons of breeder fuel a year. Reprocessing in small plants will be more costly, which must be balanced against the risk of plutonium proliferation.

Radioactive waste. Through nuclear fission a series of radioactive fission products are produced in the reactors, and through the capture of neutrons a series of radioactive transuranics. These are separated in the reprocessing and constitute a residual product which must be taken care of. Proposals have been formulated (14) for embedding most of the residual products in glass and depositing them in solid granite at a depth of 500 m. The safety of such permanent disposal schemes is now being investigated.

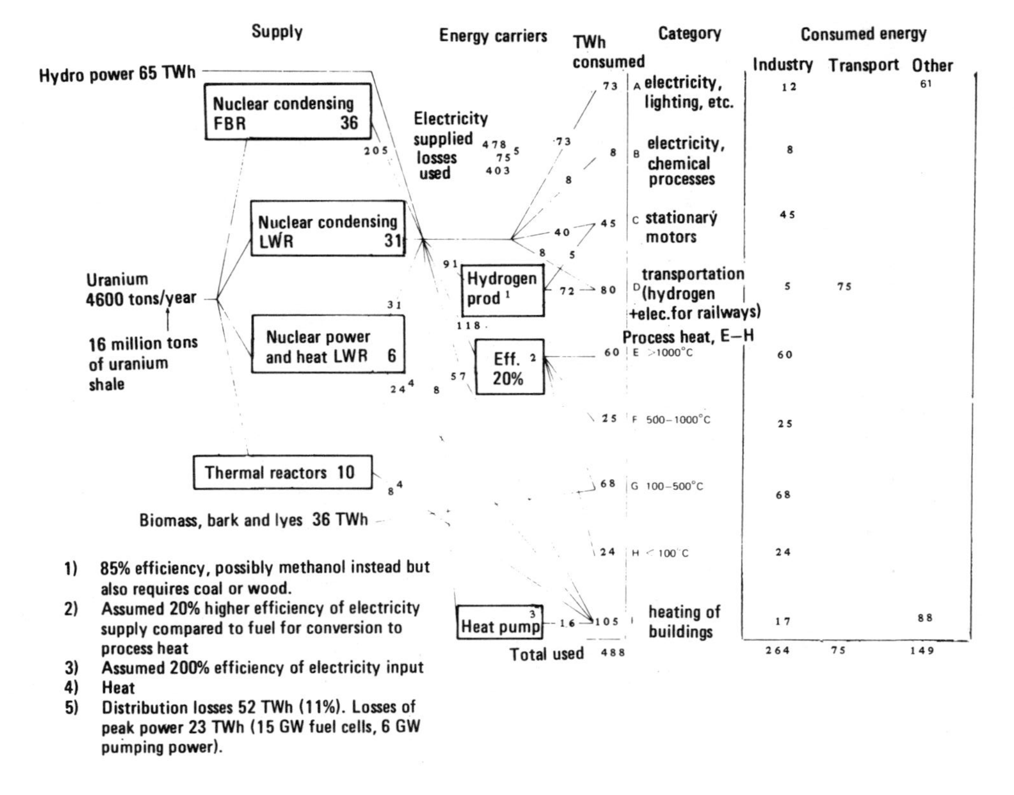

Figure 19. The energy system in Nuclear Sweden in 2015. On the left is shown the supply of energy in the system. On the right is shown the energy consumption divided into different categories A-I of energy quality. In the middle is shown the projected linkage of supply and consumption. The figures indicate the number of TWh represented by each line.

Siting, Distribution and Load Management

Locations for reactors, enrichment and reprocessing plants, and installations for storage of radioactive waste, must be designated. We shall here restrict our attention to the reactors, reckoning with the previously mentioned:

- about 67 nuclear condensing stations of 1000 MW (of electricity), including 36 fast breeder reactors
- about 6 nuclear power and heating stations of 900 MW of electricity and 1000 MW of heat
- about 10 nuclear thermal reactors of 200 MW

The six nuclear co-generation plants produce about as much heat as electricity and only a small number of localities/regions have sufficiently large heating requirements for them. Stockholm/Uppsala could be heated by, for example, two plants at Forsmark, Gothenburg by one at Ringhals, the Oresund coast by one at Barseback, together with one plant each for Vasteras-Orebro-Eskilstuna and Norrkoping-Linkoping respectively.

Nuclear district heating stations can hardly have a smaller capacity than 200 MW. That restricts the number of conceivable localities, but Boras, Falun-Borlange, Gavle-Sandviken, Jonkoping/Huskvarna, Karlstad, Lulea, Skovde, Trollhattan/Vanersborg, Umea and Ostersund appear to be possibilities.

Nuclear condensing plants will be sited along the coasts in Sweden. The choice of locations will be based on assessments of power transmission, availability of cooling-water etc. At each location 3-6 1000 MW reactors could be placed. For 73 nuclear power stations, between 12 and 24 new sites would thus be needed (CDL in 1975 reckoned with about 3 units/site) (15).

In 1972 CDL (16) made an inventory of possible sites for the 24-reactor programme then being considered and named 8 primary sites, 7 secondary sites and 3 suitable for nuclear power as well as for oil/coal (see Figure 20), giving a total of 18 sites. It is not possible without closer studies to say how many of these sites could take more than 3 reactors. If reprocessing plants are to be sited together with fast breeder reactors the site must be capable of taking about 6 reactors. This may involve the need for cooling towers. The total waste heat discharge from the nuclear reactors is about 130,000 MW, which could have consequences for the whole coastal zone (17). Knowledge of this is inadequate.

Systems for the transmission and distribution of energy, primarily electricity but also heat, need to be developed.

Regional heat pipelines may be extended and drawn to the localities which are supplied from nuclear power and heating stations. In addition district heating networks may have to be developed in localities heated by nuclear heating stations. Smaller localities with existing district heating networks could be heated by electrically operated hot water plants as an alternative to extension of electricit distribution.

The electricity network in Nuclear Sweden must be enlarged to several times its present capacity. How much larger we can only roughly estimate. At present a grid of 800 kV between the four nuclear power sites is being planned, with a total length of about 2100 km (18).

It is assumed that Nuclear Sweden will need about 5,000 km of 800 kV power lines

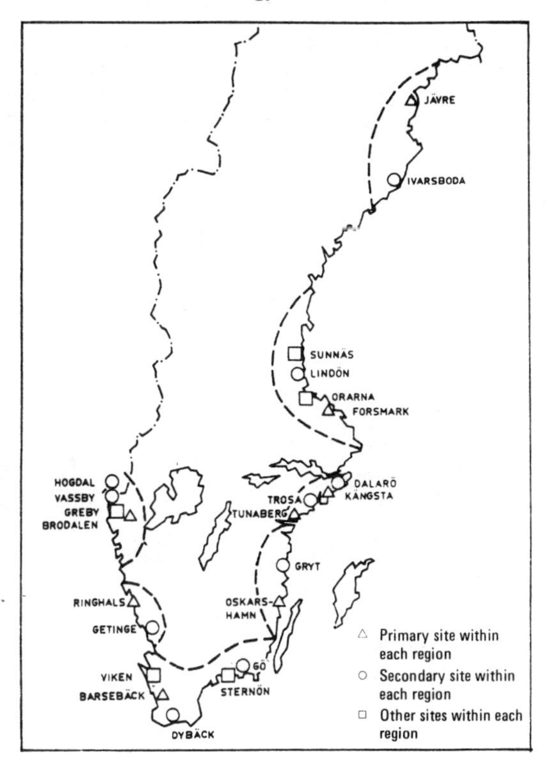

Figure 20. Main sites for nuclear power stations according to CDL's inventory (16).

(19), or possibly 12,000 km of 1500 kV. The 800 kV network would have pylons about 40 m tall at intervals of 400 metres, which gives a figure of about 13,000 pylons. In addition, the networks with lower voltage (400 kV, 220 kV, 127 kV and lower) will have to be extended.

Load management in the electricity system is a problem which must be solved. Electricity consumption varies between day and night, between weekdays, and with the seasons. As electricity is used for the heating of buildings, the variations will increase rather than diminish compared to the situation today. The variations must be taken up by other forms of power than nuclear power. Apart from hydro power one could, for instance, utilise pump storage plants (where water is pumped up into reservoirs during low-consumption periods) or fuel cells (functioning in principle as inverted electrolysis). The latter could be powered by hydrogen produced through electrolysis during periods of low electricity consumption.

We have assumed that hydrogen will be used in the transport sector. Possibly methanol may be produced instead, but that would also require supplies of coal, peat or biomass. In the transport sector in Nuclear Sweden vehicles with electric motors are chiefly used, the electric power being generated from hydrogen by fuel cells. Energy for the transport sector can be produced outside the peak load period of the electricity system and thus limit the need for pump storage plants etc.

The requirement of power for load management is estimated at around 21 GW with an assumed distribution between 6 GW pumping power stations and 15 GW fuel cells (41).

The variations can also to some extent be reduced by measures adopted on the part of users. Electrical heating with hot water storage instead of direct electric space heating will reduce the demand for peak power and is economically justified when the nuclear power component rises above a certain level (20). Water- (or air-) borne heating of buildings is thus of significance for Nuclear Sweden.

Safety

Vulnerability to interruptions, disturbances, war etc is clearly a factor which must be considered with such a heavily centralised system as that constituted by Nuclear Sweden. Hydrogen-powered fuel cells have a precautionary function in this respect apart from that of load management.

Nuclear Sweden would require 800 kV power lines with a length in the order of 5,000 km. That would imply about 13,000 pylons. (Photo: Per-Olle Stackman, Tiofoto).

Conceivable war situations or the risk of sabotage involve special requirements. The reactors must be protected so that they cannot easily be put out of action for a long period in the case of an attack. It appears to be necessary to locate an estimated three quarters of the nuclear power stations in underground rock shelters The transmission lines can scarcely be protected, but may on the other hand be relatively easy to repair.

One uncertainty is connected with the complexity of the system itself. All aspects of Nuclear Sweden are intimately connected. A failure in one area may have consequences for the whole system. These problems can be handled by overdimensioning the components. The costs of this have not, however, been included in the economic calculations given below.

Table 17. Cost estimates used for components in Nuclear Sweden (price levels at the turn of the year 1977/78) and in Swedish crowns, Mkr = million crowns. Notes to the table are printed below.

	Developed capacity	Investment costs	Operating costs
Uranium mines	5200 tons of uranium/year	1.7 MKr/year and ton[1]	260 kr/kg of natural U[2]
Enrichment (gas centrifuge plant)	3.6 M SWU	2115 kr/SWU·year[3]	300 kr/SWU[3]
Fuel element production		-[4]	
LWR		-[4]	600 kr/kg enriched U[4]
FBR			5000 kr/kg of fuel[4a]
Nuclear power stations			
Light water reactors LWR	37 of 1000 MW electricity	3500 kr/kW[6]	0.7 ore/kWh+45 kr/kW·year[5]
LWR (in rock shelter)		4375 kr/kW[7]	" ' "
Fast breeder reactors FBR	36 of 1000 MW electricity	5250 kr/kW[8]	" ' "
FBR (in rock shelter)		6500 kr/kW[7]	
Nuclear heating stations	10 of 200 MW	2000 kr/kW[8]	
Reprocessing			
LWR fuel	2 x 1100 tons	10 Mkr/year & ton[10]	0.85 Mkr/year and ton[10]
FBR fuel	360 tons	30 Mkr/year & ton[11]	2.55 Mkr/year and ton[11]
Waste handling (permanent storage)	As for reprocessing	3.1 Mkr/year & ton[12]	0.1 Mkr/ton[12]
Electricity distribution network	340[13]	0.25 kr/annual kWh[14]	- kr[15]
Hydrogen production[16]	45.5 GW	550 kr/kW	4% of annual investment[17]
Pump storage plant	6 GW	800 kr/kW[18]	" " " "
Fuel cells	15 GW	2000 kr/kW[19]	" " " "

Notes to Table 17

In the notes the following abbreviations for references are used:

EFA 2000: Energy research alternatives for Sweden in the year 2000. Energy Researc and Development Commission (DFE), 1977 (DFE reports 5,6), Liber Forlag (in Swedis

EFA URAN: Facts for EFA 2000 study, primary energy source URANIUM. Compiled by Shanker Mennon. 2nd ed. AB Stomenergi, 1977 (in Swedish).

1. According EFA 2000.
2. According to Projekt Ranstad 75, LKAB.
3. Cost estimates are lacking for gas centrifuge installations. As they are believed to involve the same cost as gas diffusion installations an estimate of the cost of the latter has been used (medium-size installation, according to EFA URAN). 90,000 separative work units (SWU) are required to produce fuel for one year for a nuclear power station of 1000 MW (of electricity).
4. Cost according to Rune Edman, State Power Board (personal communication). As the distribution between investment and operation is unclear, the cost has been treated as operating cost.

4a. Cost according to G.Vieider,AB Atomenergi (personal communication).

5. According to EKB report Summer 1977, p.6.2:15. Energy Commission
6. EFA 2000 gives 3500 kr/kW for 1980-2000. The Supply Group of the Energy Commission, EKB, gives 3020 kr/kW. Anderson-Bergendahl give the cost of Forsmark 3 as 4260 kr/kW (statement to the Energy Commission: Granskning av EKBs kistnadsmaterial, 1977-11-24.)
7. Nuclear power stations installed in rock shelters are about 25% more expensive than surface installations ("Karnkraft i berg", main report, State Power Board, June 1977).
8. As commercial fast breeder reactors do not exist at present the costs are uncertain. G. Vieider, AB Atomenergi, sets the cost of an FBR about 50% higher than that of an LWR, i.e. 5250 kr/kW (personal communication). EFA 2000 gives 4800$\pm$400 kr/kW, though this includes the first fuel core.
9. Estimate, basis lacking.
10. Costs of reprocessing are obscure at present and have risen steeply in the last few years. Nucleonics Week, December 15, 1977, gives the current price of reprocessing as $500/kg (2350 kr/kg). We have distributed this cost between investment and operation as follows. Svenke gives 6,000 million kr in investments for 25 reactors. (According to B.Svensson, L.Talmoth: "Karnbranslecykelns handelsesteg, kostnader och problem". Stockholm University, Institute of Business Economics, June 1977). This gives an investment of 10 Mkr per year and ton. Written off over 15 years and at 15% interest the annual capital cost will be 1500 kr/kg. The difference (2350-1500) of 850 kr/kg is taken to be the operating cost. A life span of 15 years is assumed. The AKA enquiry (SOU 1976:31, Spent nuclear fuel and radioactive waste, English translation) estimated the costs in 1976 of a reprocessing plant at 3.75 Mkr/year and ton.
11. G. Vieider, AB Atomenergi, writes in EFA URAN that the specific costs of reprocessing breeder fuel can be estimated to be 2-4 times higher than for LWR fuel. The fast breeder reactor core is assumed to be 10 tons and the uranium blanket 5 tons (which is reprocessed as light water reactor fuel).
12. Cost very uncertain. Tons per year calculated as for reprocessing. EFA 2000 annex PROD gives investment as $500 M and operation as $16 M/year for 750 tons per year. Equivalent to 3.1 Mkr/ton per year in investment, 0.1 Mkr/ton in operation.
13. 478 TWh of electricity are produced. The electricity network is taken to be already developed for production of about 140 TWh (hydro power plus power equivalent to the nuclear power programme with 13 reactors).
14. Stated by the Supply Group of the Energy Commission (report Summer 1977). Basis deficient. In a study from the Office of Technology Assessment, USA, "Applica-

tion of Solar Technology to Today's Energy Needs", the distribution and transmission costs are given as $0.75 per $ of installed productive capacity, which is considerably higher than the figure used here.

15. Information lacking
16. Hydrogen production is assumed to take place at low load, involving short average usage times. Assuming 2000 hours, 91 TWh of electricity consumed, gives a capacity of 45.5 GW. EFA URAN states 550 kr/kW as fixed costs.
17. Estimate, not including cost of utilised electricity.
18. Cost according to EFA 2000, annex PROD, with subterranean store.
19. Cost according to EFA 2000, annex PROD.

What would Nuclear-Sweden Cost?

We shall here make a very rough calculation of the costs of Nuclear Sweden and also look at how much energy would be supplied in different years. This is a difficult task, not least because the cost estimates have rapidly changed in recent years. The figures used in the calculations are presented in Table 17 and the result is shown in Figure 21 and Table 18.

The costs will reach a peak around the year 2010. This is because it takes a relatively long time to construct nuclear power stations etc. For that reason some years will elapse between capital outlays and energy production; furthermore, only the costs of developing the system to the year 2015 have been included. Investments for installations which are to come into operation after 2015 have thus been omitted. After the establishment of Nuclear Sweden the main investments required, if energy demand does not increase, will be for replacement of worn out installations.

Table 18. Costs for energy production 1990-2015 in Nuclear Sweden in thousand million kronor.

	Investment	Operating costs
Uranium mines	9	26
Enrichment	8	20
Fuel production	-	22
Nuclear power stations		
LWR	154	48
FBR	223	19
Nuclear heating stations	4	-
District heating network, regional	4	
Reprocessing		
LWR	22	13
FBR	11	6
Permanent storage of radioactive waste	8	2
Electricity distribution network	85	-
Hydrogen production	25	11
Peak power (pumping power plus fuel cells)	35	15
	588	182
	Total about 800 thousand million kronor	

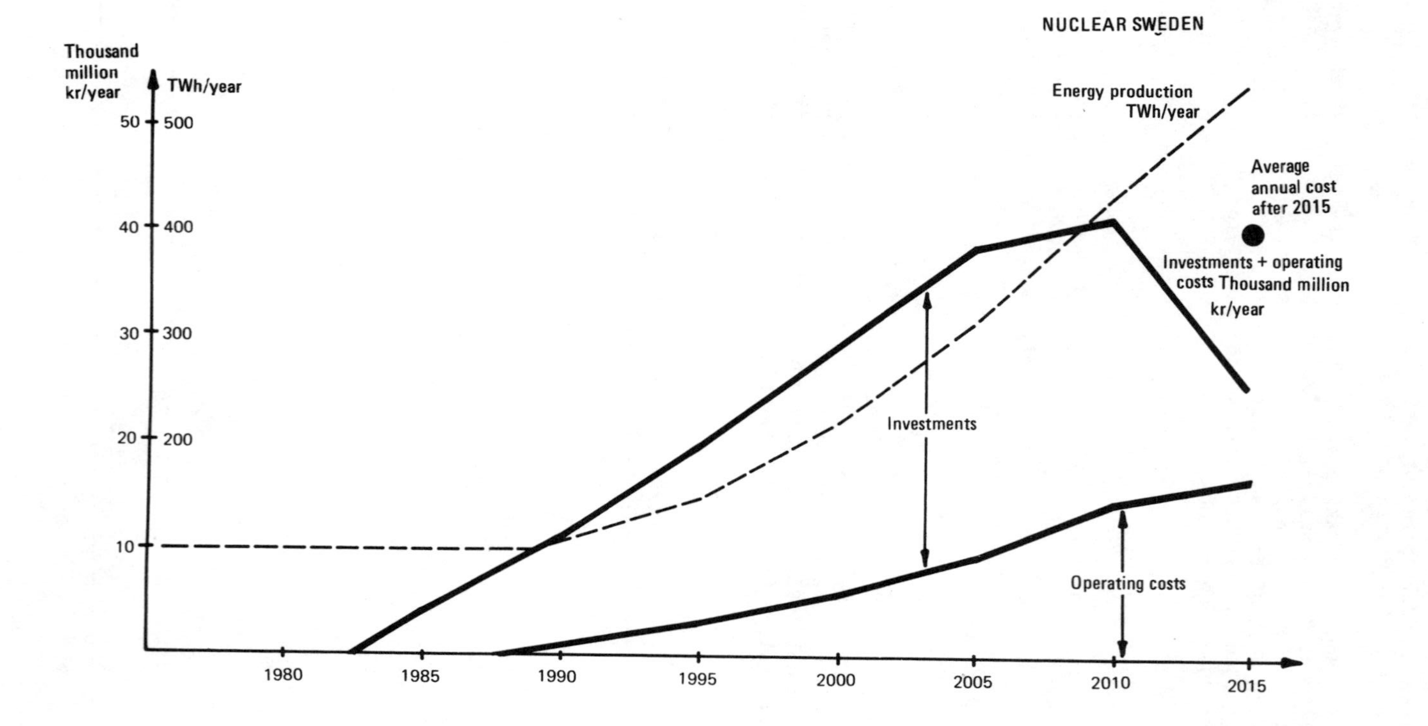

Figure 21. The cost and production of the energy system in Nuclear Sweden. The fall in investments after the year 2010 is due to the omission from the calculation of the replacement investments which will then become necessary for installations which are to operate after the year 2015. If these are included the investment need will decline somewhat for about a decade before increasing again later, when the large investments around the year 2000 must be renewed. An average cost for the system after 2015, calculated over such periods of time, is shown in the diagram.

Some uncertainties and obscurities concerning Nuclear Sweden

There are a number of uncertainties and obscurities concerning Nuclear Sweden. One of them is the uncertainty of technical performance. To these uncertainties belong the doubling time of the breeder reactor, how reprocessing of breeder fuel should be organized etc. (Compare the section "Nuclear power" in chapter 3).

Another uncertainty concerns the economics of this model. The costs of the light water reactor have increased rapidly, calculated both in fixed prices and in physical measures such as the number of working hours per installed capacity, as shown in Figure 22. Various elements in the system have also been affected by cost increases. Reprocessing is perhaps the most striking case, the reason being presumably that the development of the technology has not yet been completed for the high rate of burn-up which the light water reactors have involved. The even higher rate of burn-up of the breeder reactor fuel increases the uncertainty still further, and the costs of breeder reactors are therefore very uncertain. Taken together, several of the uncertainties indicate that the costs for Nuclear Sweden may have been underestimated.

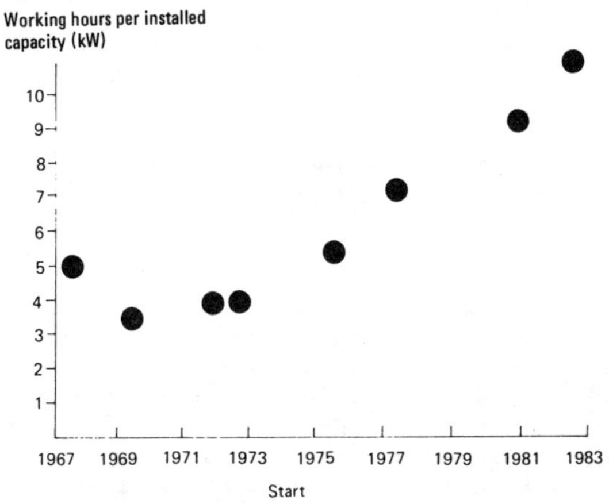

Figure 22. Changes in demand for labour power for construction of nuclear power stations in the United States (21, 22).

The environmental uncertainties include the safety of both the light water and the breeder reactors. Reactor accidents, e.g. core melt-downs, can lead to extensive discharges of radioactive substances.(1) The likelihood of such accidents is under debate (12). Another uncertainty is the working environment in reprocessing plants (see Nuclear power in chapter 3). In addition there are questions in connection with the mining and recovery or uranium. How the very large amounts of cooling water (130,000 MW) will affect the environment is also uncertain.

Another problem is the alteration of the landscape brough about by the pylons (10-

(1) Provided that the probability of a core melt-down remains constant i.e. not affected to a decisive degree by aging phenomena or technical developments, it is possible to estimate the probability of a core melt-down during 30 years of operation of 75 reactors (12). If one assumes, with the Rasmussen report (42), that the probability per reactor is one in twenty thousand reactor years, the result will be a 10% risk of a core melt-down. Rasmussen indicates a margin of uncertainty of ± a factor of 10. At the upper limit there is then a 65% risk and at the lower a 1% risk. A core melt-down will not necessarily cause great damage to the environment, and in Nuclear Sweden that is also counter acted by the location in rock.

15,000, 40 m high) of the main transmission grid. The land requirements for the energy system in Nuclear Sweden are relatively limited; the power-lanes are the major item here. The location of the reactors will be determined by the technical demands on their siting.

The political disunity about the future of nuclear power in Sweden is a serious problem if oil has to be replaced by nuclear power within 35 years. That task requires a concentration of forces and political consensus of a kind which is uncommon in peacetime. The great complexity of the nergy system of Nuclear Sweden demands a high degree of co-ordination and thus of centralised decision-making. Plans for the siting of reactors and the components of the fuel cycle cannot be allowed to be overturned by local authority vetoes, and the legislation needs to be adapted for this purpose.

A decision to introduce Nuclear Sweden will involve the allocation of resources in good time to develop uranium mines, enrichment and reprocessing. The development of the education system will also be affected by the change of energy system. In Nuclear Sweden about 20,000 people will be needed to operate the reactors and many of these must be highly qualified technicians. Many more specialists on radioactive environments will also be needed. In the section "Similarities and dissimilarities between Nuclear and Solar Sweden" we shall discuss the socio-economic aspects.

The risk of plutonium going astray is another important question which does not relate so much to technology but rather to the organization, etc., of society.

Solar Sweden

The same basic preconditions apply to Solar Sweden as to Nuclear Sweden, including a doubling of industrial production and an energy consumption of 488 TWh by the year 2015. See further p.67. A detailed description is presented in a previous report "Solar Sweden"(1), and this section is a summary of that report.

The idea of basing a very large part of Sweden's energy supply on renewable energy sources is a new one. The energy commissions of the last few years appointed by the state and industry have come to the conclusion, after more or less thorough studies, that the renewable energy sources can only play a marginal role in the future. In this study we have arrived at the opposite result: our assessment is that an energy system based entirely on renewable energy sources would appear to be possible in Sweden in 2015.

We have proceeded from existing technology and from the techniques which are at present somewhere between the laboratory and demonstration stages. The basic principles have been demonstrated and certain additional experience has been gained. The ways of collecting energy are thus basically the following:

A. Energy collected directly from solar radiation
   1. by means of collectors
   2. by means of solar cells

B. Energy collected indirectly from solar radiation
   1. through biological systems
   - energy forest
   - marine biomass (algae etc)
   - straw, reeds, forest waste

(1) Available from the Secretariat for Futures Studies, PO Box 7502, S-103 93 Stoc

2. through reophysical systems
- hydro power
- wind energy

The energy is collected principally in the form of electricity (from hydro power, wind power, solar cells), of hot water (from solar collectors) and of solid fuels (biomass from energy forests, straw, reeds, forest waste etc). It is partly converted into (secondary) energy carriers and is delivered to the end users. The choice of secondary energy carriers has been made after a plausibility estimate in which the guiding principles have been

to limit conversion between different forms of energy (i.e. utilise biomass directly, without conversion to "refined" fuels or electricity which involves considerable conversion losses)

to utilise the waste heat which occurs when conversion is necessary (from co-generation, fuel cells or methanol production)

that energy generation should take place as close to the users as possible.

Methanol is the only new energy carrier to be introduced and that will be used chiefly as motor fuel.

It should be emphasised that the energy system which we have outlined here is neither the only conceivable one nor necessarily the best. It is, on the other hand, a possible system and more profound studies will have to indicate what improvements can be made.

The whole systems is shown in Figure 23. The supply to the various areas of utilisation is presented in Table 19.

Table 19. The energy supply of Solar Sweden in 2015 distributed between traditional user sectors and kinds of energy. TWh/year.

Industry		
Electricity	71	
Methanol	10	
Biomass*	86	
Other fuel (bark and lyes)	36	
District Heating**	44	
Solar heating	17	
	264	264
Transport sector		
Methanol (vehicles)	67	
Electricity (railways)	8	
	75	75
Housing, services etc.		
Heating of premises		
Co-generation	21	
Solar heating	54	
Back-up heat (biomass, electricity)	13	
Electricity (lighting etc)	61	
	149	149

* Excluding bark and lyes
** Waste heat from fuel cells and methanol production

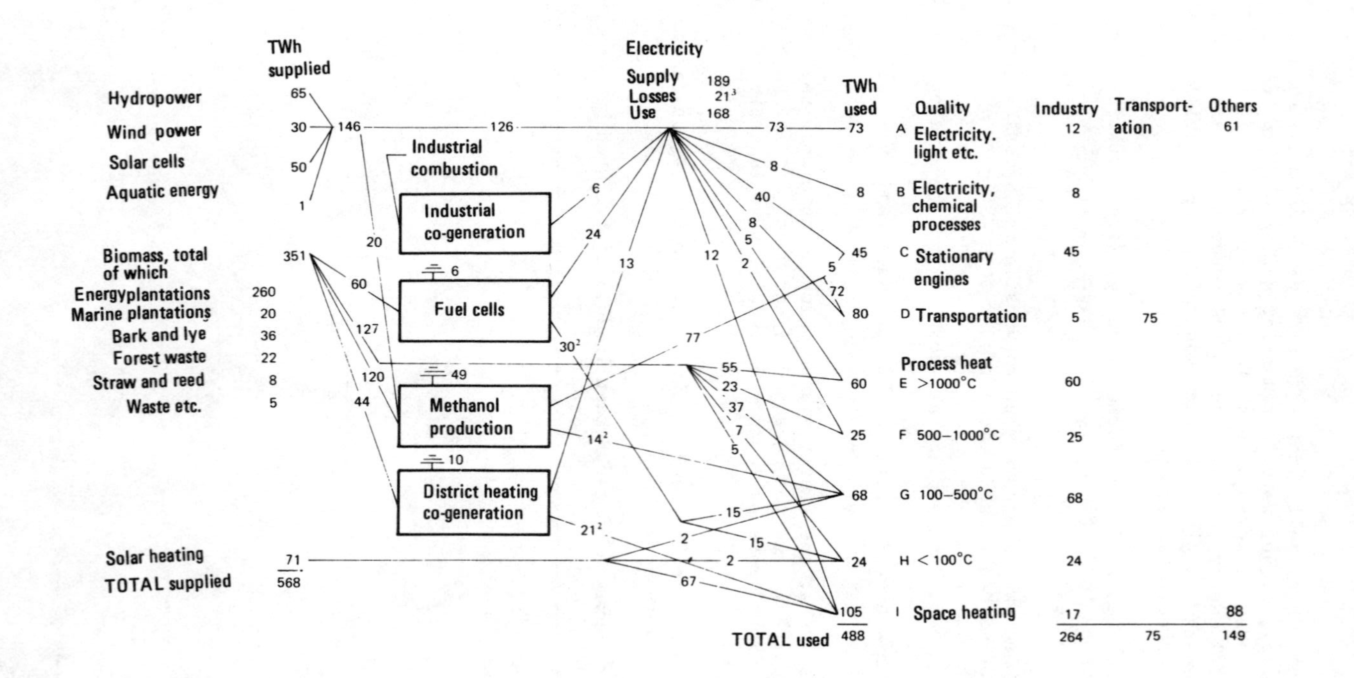

TWh used	Quality	Industry	Transport-ation	Others
73	A Electricity. light etc.	12		61
8	B Electricity, chemical processes	8		
45	C Stationary engines	45		
80	D Transportation	5	75	
	Process heat			
60	E >1000°C	60		
25	F 500–1000°C	25		
68	G 100–500°C	68		
24	H < 100°C	24		
105	I Space heating	17		88
488 (TOTAL used)		264	75	149

Figure 23. Energy system 2015. On the left is shown the supply of renewable energy in the system as well as the amount contributed by the various techniques. On the right is shown the energy consumption distributed between the various categories of energy A-I. The diagram in the middle shows the planned linkage between supply and consumption. The figures indicate the number of TWh represented by each line.

The Energy Supply of Solar Sweden

Energy forests consisting of fast-growing species of trees (poplar, sallow, alder, birch) make the greatest contribution to the energy supply. Special demands are made on the soil, which must be rich in nutrients and well watered. Our calculations are based on an average yield of 90 MWh per hectare and year, which is half of the maximum yield measured in experimental plantations.

Table 20 shows the relationship between forest and swamp land in Sweden and how this can be used for the production of biomass (23).

In addition certain marginal lands, e.g. abandoned arable land, power lanes and coastal land, can be utilised. This could provide several 100,000 ha more.

Table 20. Land use for production of biomass, millions of hectares.

	Total area (Mha)	Utilised for energy forests (Mha)
Forest land	24	1.2
Swamp land	5	1.7
Sweden's total area	41	2.9

Total energy from energy forests of 2.9 Mha will be 261 TWh, which represents about 87 M tons of wood (with 40% moisture after self-drying). The area is about 6-7% of the total land area of Sweden and somewhat less than the agricultural area (9%). The area of energy forests broadly equals that of arable land reclaimed from the beginning of the nineteenth century to the 1940's.

There are naturally numerous questions which have to be studied further. The possible ecological disruptions are an example. Another is that somewhat less land will be available for timber growing.

Cultivation of algae, etc, in ponds, lakes or bays of the sea can provide high yields of biomass. The cultivation can be combined with sewage purification plants and the utilisation of waste heat to increase efficiency (200-400 MWh/ha and year). The algae can be converted into methane gas through anaerobic decomposition (absence of oxygen). It is possible that this type of biomass cultivation will turn out to be more attractive than energy cultivation on land. We have assumed here that energy amounting to 20 TWh/year can be produced, which corresponds to an area of about 100,000 ha at 200 MWh/ha and year.

Straw which is produced in agriculture can be simply gathered and incorporated in the energy system. If 20% is utilised, that would represent about 5 TWh/year.

Reeds are common in Sweden and the dry-substance production amounts to 25-50 MW/ha and year. Techniques for harvesting etc have been developed in lake restoration projects. We assume that 60,000 ha can be utilised and this gives 3 TWh/year.

To collect 4.5 Mtons of forest waste is regarded as feasible, which gives 22 TWh/year.

The timber industry consumes about 36 TWh annually from bark and lyes. We assume that this figure will still apply in 35 years' time.

The annual amount of waste has an energy content at present corresponding to about 40 TWh. Of this amount a part is reused at present as raw materials. The energy contribution from waste may be estimated at 5 TWh per year.

Energy forest. Solar energy is absorbed when the green plants grow. Plantations of quick-growing deciduous trees can bind up to 200 MWh/hectare (1 hectare = 10,000 m^2). A corresponding value for coniferous forest on good forest land (whole tree) is 25 MWh/ha. (Photo: Pressens Bild).

Solar heat could play a large role in the heating of buildings, which only requires a low intensity (W/m^2). The techniques we have for collecting solar energy for low-temperature heat are also relatively efficient (about 30%).

One difficulty is that the Solar heat must be stored from the warm half-year to the coled half-year (cf. Figure 25). Several development projects exist in this field in Sweden and abroad which, apart from the design of solar collectors, are also concerned with storage, e.g. in the soil, in large water tanks, in salt smelters (where latent heat is utilised), in distillation storage plants etc.

With large heat stores (corresponding to the energy needs of a few dozen houses) the heat losses can be limited, and thereby the cost per amount of stored energy. We have calculated that energy corresponding to 71 TWh can be obtained from solar heat. Complementary heating is also needed to a certain extent.

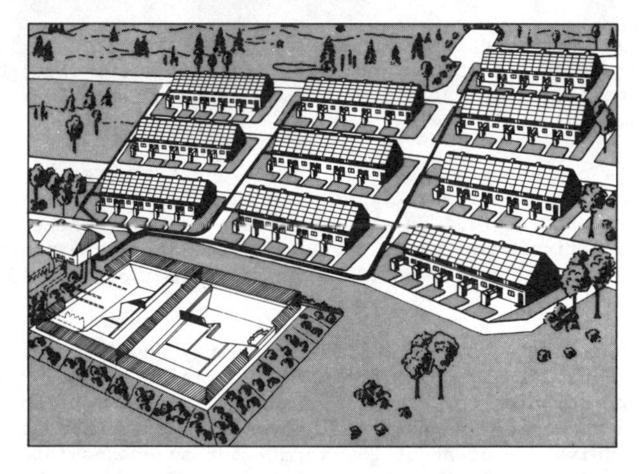

Figure 24. In Linkoping a private construction firm is carrying out a project with solar energy houses with support from the State Council for Building Research on the lines of this sketch. The houses are connected to a collective storage system, built as a covered pond.

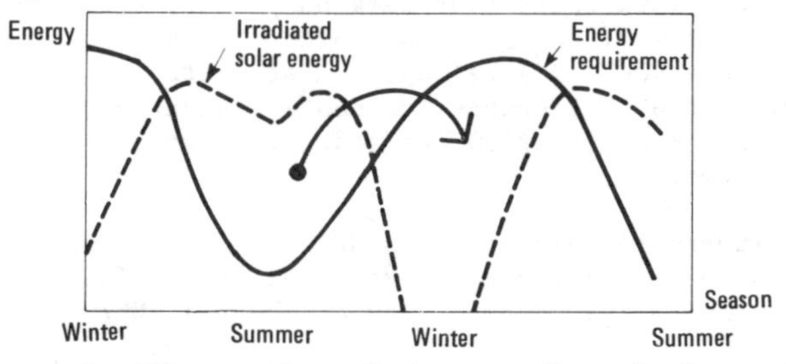

Figure 25. Diagrammatic illustration of the annual variations of solar radiation in relation to the heat requirement of a Building at Swedish latitudes. Both the requirement and the possibility of seasonal storage are indicated (24).

<u>Hydro power</u> in Sweden could, on a purely technical assessment, be developed to an annual production of 130 TWh, of which 95 TWh is economically feasible at today's energy prices. The decision of Parliament on energy policy in 1975 involved the production of 66 TWh a year by 1985. We have assumed an energy contribution of 65 TWh a year from hydro power.

<u>Wind power</u>. The basic principles of wind power generators (25) have been known for decades and several million such generators have previously been utilised. The technical development of wind power has now been taken up again, by the aircraft industry among others.

Sweden lies within the so-called West-wind belt and therefore has favourable wind conditions. From an economic viewpoint the siting of wind power generators in windy areas is desirable, which in practice means the coasts, Oland, Gotland, the Smaland plateau, Lake Vanern, Uppland, Lake Vattern and the Ostergotland plain. A total of 30 TWh/year is assumed in our example, This energy is generated by 3700 wind power stations with a capacity of 4 MW (with a tower 100 m tall and a propeller diameter of about 100 m). On average a wind power generator produces energy during 2000 hours a year (1700-2500 depending on the location). Added to these are small wind power generators of 8-40 kW, mainly for local needs (electricity or heating) for agriculture, summer cottages, etc.

Solar cells (e.g. photovoltaic cells) convert sunlight directly into low voltage direct current. By series connection of several cells the voltage can be raised, converted to alternating voltage, transformed up to a suitable voltage level and connected to the electricity mains.

Several materials can be used, but silicon cells are at present regarded as those for which the costs can most quickly be reduced. Silicon cells have today reached an efficiency of more than 12% in the conversion of solar energy to electrical energy. The theoretical maximum lies around 20% and it is expected that they can yield 15% in large-scale operation (26). They yield about 100-150 kWh per year and m^2 in Sweden.

In order that solar cells may play a role in energy supply, their capital costs must be reduced considerably. In 1975 the cost was of the order of 50 kr per watt at maximal solar radiation (peak watt). The US Department of Energy, (DoE) has established as a goal for 1986 that silicon cells should cost about 2.5 kr per peak watt (the equivalent of about 250 kr/m^2) for an encapsulated cell (27). So far development has moved faster than expected and it is possible that the goal will be reached before 1986.

The solar cells are assumed to be mounted on roofs and walls of or close to existing houses. They can form roofs over parking places, etc. No central solar power stations have been assumed. An area equivalent to about one tenth of all urban land (at present about 4000 km^2) is regarded as suitable for solar cells (possibly in combination with solar collectors for heat). We have calculated with an energy contribution from solar cells of 50 TWh/year, corresponding to about 400 km^2, i.e. 50 m^2 per inhabitant.

Aquatic energy. There are possibilities of extracting energy from sea waves (28) and from salt gradients (29) (by the mixture of water of differing salinity, e.g. at river mouths). An energy contribution of about 1 TWh/year has been assumed here.

Construction of the Energy System

As shown in Figure 23 the energy sources are connected together in different sub systems. The biomass is used directly in the (processing) industry, as fuel for co-generation plants, and as a raw material for the production of methanol for the transport sector.

The fuel cell - as we have already mentioned in the section on Nuclear Sweden - is in principle an inverted electrolytic cell, i.e. it converts oxygen and hydrogen into electricity (and water). The modules are small. Larger installations are obtained by linking together small modules. Of the energy supplied 40% is converted into electricity and 50% into heat in the temperature range 70-165°C. Their power output can easily be varied, so that fuel cells can be used to equalise load fluctuations. The effects on the environment are very small.

A new energy carrier must be introduced in Solar Sweden, chiefly to replace petrol in the transport sector. Several reasons suggest the choice of methanol. It is amenable to storage, is fluid and can be successively introduced in existing vehicles. Hydrogen gas, which has often been discussed, would demand new technological development without the same possibility of a gradual transition. Methan gas involves the same difficulties. Methanol has half of the energy density of petrol (per unit of volume).

Methanol can be produced from biomass, peat, natural gas or coal. It can be used in several ways, e.g. in combustion engines of the present-day type. Methanol is also a usable fuel for fuel cells. The transport sector in Solar Sweden will be run on methanol-powered fuel cells. The electricity produced will drive electric motors. In that way the efficiency in the use of fuel will increase to twice the kilometerage per kWh as compares with the petrol engines of today.

Methanol can be mixed with petrol up to 15-20% without significant alterations to the present car engines. With somewhat greater alterations petrol engines can be run on pure methanol. The production and distribution of methanol can therefore be developed early. The admixture of methanol means that the lead in petrol can be dispensed with altogether, and other pollutants will also be reduced.

The surplus electricity from solar cells (during the warm half of the year) can be transferred to methanol production in the form of hydrogen and oxygen through electrolysis of water. This will lead to higher efficiency in the output of methanol. An average efficiency of 55% has been assumed in the conversion from biomass to methanol. Part of the waste heat from methanol production has been utilised in industry.

Electricity System

In Solar Sweden electric power is upplied by many different installations (see Table 21). These are linked together into one system. The availability of power from the installations will vary, in the case of wind power and solar cells, with the prevailing wind force and solar radiation. The variations can therefore be both rapid and large. This is counteracted to some extent by the dispersal of the installations over a large part of the country. When a surplus is generated, this must obviously be stored for later used (see below).

Electricity is generated by hydro power, wind power, solar cells, aquatic energy and from co-generation, industrial back pressure and fuel cells.

Table 21. Electricity production system of Solar Sweden.

	Energy per year (TWh)	Installed capacity (GW)
Hydro power	65	14
Wind power	30	15
Solar cells	50	(c.50)
Aquatic energy	1	
Co-generation	13	3
Industrial back pressure	6	1
Fuel cells	24	c. 8
Pumping power stations (or equivalent)	0	c.15
Total electricity production	189	

A production of 13 TWh per year has been assumed for co-generation plants, corresponding to an installation of about 3000 MW (electricity). The co-generation plants can be based on steam turbines or diesel engines (fired with biomass, e.g. in the form of pulverised wood). The co-generation plants can be designed with hot water accumulators and therefore contributed to the load management in the electricity supply system.

We have assumed that 6 TWh of electric energy can be produced in back pressure installations in industry, which is in accordance with SIND's forecast (2).

Load Management

The production in Solar Sweden of electric energy with renewable energy sources places special demands on the load distribution. The demand must be balanced against the supply, and surplus capacity must be stored. How this regulation is to take place has not yet been investigated in detail, butit can in principle be done as follows.

Short-term Regulation (day-week)

Hydro power. By expanding the capacity, joint operation with 5-7 GW wind power is throught to be possible. The existing hydro power can be interconnected with 3-5 GW wind power (31).

Fuel cells can rapidly vary their output and thus compensate for the variation in output from solar cells and wind power.

Co-generation. Through thermal stores the daily variations in electricity demand can be partly balanced.

Electric heating with hot water storage in parts of the total stock of buildings; the back-up heat output can for instance be allocated to low-load periods or periods when a surplus of electricity is produced. Control (switching on and off) can be effected through radio frequency signals over the mains.

Regulation of demand in industry. In processes where the processing heat is generated with electricity the electrical load can be affected by utilising thermal stores or by producing hydrogen by electrolysis of water. The hydrogen gas is stored when required.

Apart from these ways of regulating demand there will probably be a need for installations where the electric energy can be stored for a relatively short time, e.g. pump storage plants. In these the electric energy is transferred to potential energy of water by pumping the water up into a high-lying lake or dam. The electric energy is subsequently produced as required in the form of hydro power. Another possible alternative would be mechanical fly-wheels.

Seasonal Storage

The need for seasonal storage (summer-winter) applies both to solar heat and to energy from solar cells and hydro power. At most 35-40% of the electrical energy from solar cells needs to be stored from sunny summer months to less sunny winter months. One alternative is to supply oxygen and hydrogen, from electrolysed water, for the production of methanol and thereby connect the solar cell electricity to the energy system in the form of methanol. 20 TWh/year of the energy produced by the solar cells are assumed to be utilised in this way.

The load fluctuations are thus counteracted by measures both on the user and production side. A large part of the electricity production also takes place near the load. These two factors mean that the demand on the capacity of the mains is limited.

The supply of electricity in Solar Sweden will considerably exceed today's level. The reason for this is that the renewable production methods will provide electricity from hydro power, wind power and solar cells. Another reason is that part of the heat requirement, mainly district heating in the large cities and part of the processing heat, cannot be provided by solar heat but is produced in installations with back pressure power, which further increases the electricity supply.

Rates of Introduction

Data for a discussion of rates of introduction are largely lacking. We have assumed that the energy system will have been introduced around the year 2015, i.e. in just over 35 years. On that basis an assessment can be made as to whether or not it will be possible to develop the system so quickly.

We have assumed that production of the installations takes place in one year and that these begin to produce energy the following year. For biomass a delay of two years has been assumed. The assumed rates of introduction are shown in Figure 26. The times of introduction are set later than in estimates by the National Swedish Board for Energy Source Development (NE) and the rates of introduction are also initially lower than those assumed by NE (32). We have deliberately adopted a fairly cautious level.

Contributions to energy supply grow as shown in Figure 27. The expansion of the renewable energy system will be barely 6% per year over the period 1990-2015. Today the share of energy from renewable energy sources constitutes about 24%, by the year 2000 it will have increased to about 50%, and will amount to 100% around the year 2015.

A growing use of biomass can begin immediately. Forest waste, straw, reeds etc are available and can now be put ot use. Brushwood grows in many forest areas and other areas can be used for such extensive cultivation of biomass. Peat extraction could constitute another point of departure. In a later phase more extensive forms of cultivation can be introduced. From 1990 onward 100,000 ha per year will be cleared for intensive energy forest production. By way of comparison it may be mentioned that forest planting during 1973 embraced 175,000 ha and that the arable area decreased during the period 1956 to 1971 by, on average, 30,000 ha per year.

For solar heat the preconditions also exist for an early start. Installations for solar heated tap-water are already being produced in the country. The market can be stimulated by the state through loan regulations, etc. Other possibilities of utilising solar heat in an introductory phase are for the heating of swimming pools drying of grain, etc. Other components are at different stages of research, development or demonstration.

Costs of Solar Sweden

The work of research and development of renewable energy forms only gathered momentum after the oil crisis of 1973-74. The techniques have therefore not been fully developed. This naturally makes all cost estimates very uncertain. For Solar Sweden we have made use up to 2015 of cost estimates which are generally regarded as valid for the 1980's. There is reason to assume that this is a pessimistic estimate. Solar Sweden may therefore very well be cheaper than indicated here.

Within many fields an empirical curve has historically been followed where the cost per manufactured unit falls by a certain percentage for every doubling of the volume. For other systems this trend has not applied. The costs of nuclear power, for example, have instead risen, as was shown in Figure 22. An important difference between large energy production plants and renewable energy technologies is that the latter, thanks to their small units, are probably better suited to industrial mass production and it is in that context that cost reductions can be expected. In the cost estimates, however, this expected cost reduction has not been taken into account.

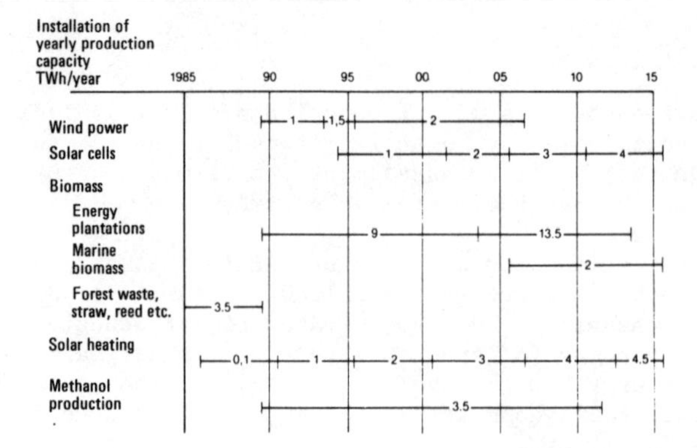

Figure 26. Assumed rates of introduction for different kinds of energy. The figures within the diagram relate to the annual installation of energy production capacity expressed in TWh/year.

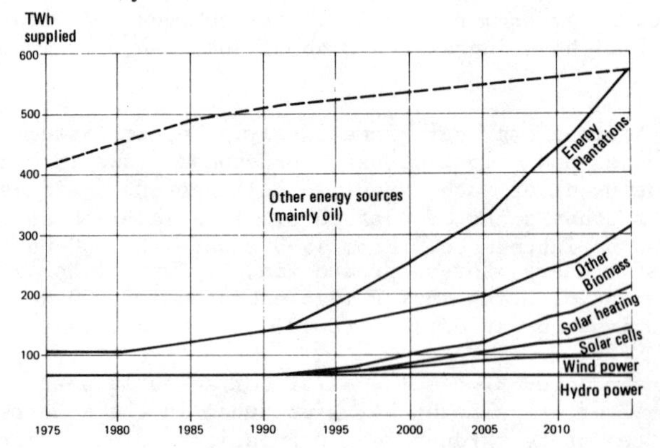

Figure 27. Contributions of renewable energy forms to energy supply in Solar Sweden. The broken line shows the assumed development of energy demand. It is close to SIND's forecast (alternative A) up to 1995 and thereafter increases more slowly.

By means of the cost estimates given in Table 22 and the rates of introduction discussed above we have calculated how the costs and energy supply of Solar Sweden will increase between 1990 and 2015. As shown by Table 23 the cost of solar cells is a predominant part. If one changes the energy system so that the 50 TWh of electricity provided by the solar cells are instead produced with fuel cells using biomass as fuel, the total costs will fall by 140 thousand million kronor (almost 20%). If instead of the pulp wood price for energy forests we use the cost estimates which have been published (cf. Table 22, note 30), this means that the total costs will fall by 90 thousand million kronor (about 10%).

Uncertainties

The uncertainties in the arithmetical example presented here are naturally very large. This is true both of the use of energy, how it is to be distributed between different supply possibilities, and the development of various components. Let us briefly describe the possible variations.

The need for land is influenced by the yield from energy plantations. If we assume that this amounts to 120, instead of 90, MWh per hectare and year we find that the area requirement will be reduced to 2.2 Mha. If we assume that productivity is reduced from 90 to 70 MWh per hectare and year, the land requirement will increase from 2.9 to 3.7 Mha. For production, fertilising and watering an energy supply of 15 MWh/ha and year is required. This figure, too, can vary and thus affect the final result.

Table 22. Costs of utilising renewable energy techniques. The notes to the table are printed below.

	Installed capacity by 2015	Investment Costs	Operating costs
Wind power	15 GW	4000 kr/kW[1]	2% of annual investment*[7]
Solar cells	(50 GW) 500 km^2	c. 2.5 kr/W (peak)[2] plus 100 kr/m^2 installation, total 400 kr/m^2	2% of annual investment*[8]
Solar heating	71 TWh/year	2kr/annual kWh[3]	2% of annual investment*[9]
Methanol production	16 Mton (77 TWh)	1100 Mkr/750,000 tons per year[4]	4% of annual investment*10
Pump storage plant	15 GW	800 kr/kW[5]	4% of annual investment*10
Fuel cells	8 GW	2000 kr/kW[6]	4% of annual investment*10
Biomass			
Energy forest plantation	260 TWh/year		300 kr/ton dry weight[11]
Other biomass (excl. bark and lyes)	55 TWh/year		230 kr/ton dry weight[12]

* For maintenance and service

1. The National Swedish Board for Energy Source Development (NE) suggests 3400-4500 kr/kW for 4 MW prototypes. Cf. Solar Sweden.
2. 2.5 kr/watt (peak output) corresponding to ERDA's goal for 1985, cf. Solar Sweden.
3. The Building Research Council suggests 1-6 kr/annual kWh. The study by AB Atomenergi suggests 0.6 kr/annual kWh (peter Margen: "Kombinerat magasin och solfangare for grupphusbebyggelse". AB Atomenergi, duplicated, February 1977.
4. According to Group B of the Energy Commission, report in Summer 1977.
5. See note 11, table 17.
6. See note 12.
7. NE suggests about 0.5%
8. We have used the same figure as for sun-collectors, see note 9.
9. Given by D Hirsch, ERDA, before the Subcommittee of the Committee on Government Operations of the US House of Representatives, published in "New Developments in Solar Energy", February 1, 1977. USGPO.
10. Assumed values.
11. The costs of biomass have been estimated in various studies. The commercial price of ordinary wood (for the pulp industry) is about 300-320 kr/ton DS (dry substance) at a motor road (I). This represents 6 ore/kWh if the wood is burned. For energy forest plantations various sources give:
    a) 80 kr/ton DS: 2 ore/kWht (II)
    b) 150 kr/ton DS: 3 ore/kWht (III)

c) 2-3.3 ore/kWh at a production of 67-175 MWht/ha (IV)
d) 1-2 ore/kWt (V)
These figures should be used with caution as the methodology of the cost estimates is not fully known. For comparison the price of heating oil can be given, 5-6 ore/kWh. We have used 300 kr/ton DS.
I. Memorandum by P O Nilsson and others "Preliminar bedomning av skogen som energikalla". Royal College of Forestry, Stockholm, 1976-12-09.
II. EFA 2000, annex PROD, p. 9:2: "Wood chips produced in a central Swedish chip chain cost, when delivered to the roadside, 80 kr/ton dry substance. Ther is reason to believe that a rationally conducted energy cultivation should give a lower production cost than the chip chain".
III. Olle Lindstrom: Preliminary assessment, variation between 50 and 250 kr/ ton DS. Quoted in Energy Commission, Group B, report summer 1977, p. 4.8:9.
IV. Silvicultural Biomass Farms, Vol IV, p. 5-10, Mitre Technical Report No. 7347, Mitre Corp., USA, May 1977.
V. M D Fraser, I F Henry and C W Vail: "Design, Operation and Economics of the Energy Plantation". Intertechnology Corporation, Jan. 1977, USA.
12. Ref. I in note 11.

Another factor which can palpably affect the requirement for land for energy cultivation is the development of marine cultivation of biomass. In our example we have assumed a supply of 20 TWh/year from marine cultures. The availability of sea and lakes is considerable in Sweden. This makes marine cultures very attractive.

There are possibilities of utilising the biomass from agriculture and forestry more effectively. We have calculated with 30 TWh/year, but it is possible to collect at least twice that amount provided that it is done with suitable techniques.

Table 23. Costs in thousand million kronor for Solar Sweden and its energy production 1990-2015

	Investments	Operation
Solar cells	200	28
Wind power	60	20
Solar heat	142	26
Methanol production	24	14
Pumping power	12	7
Fuel cells	16	5
Co-generation	10	2
Biomass		
Energy forest plantations		177
Other biomass		47
	464	326
	Total c. 800 thousand million kronor	

In Sweden there is a possibility of a considerably larger energy production from wind power gnerators than the 30 TWh/year which we have included.

It is also conceivable that other techniques may become of interest. An example might be methods for producing electricity from small temperature differences (so-called heat engines). According to certain evidence the costs of such engines would appear to be so low that they could more than outweigh the drawbacks of low efficiency. They would make it possible to produce electricity also from solar heat installations. They could also possibly be combined with heat pumps (27, 33).

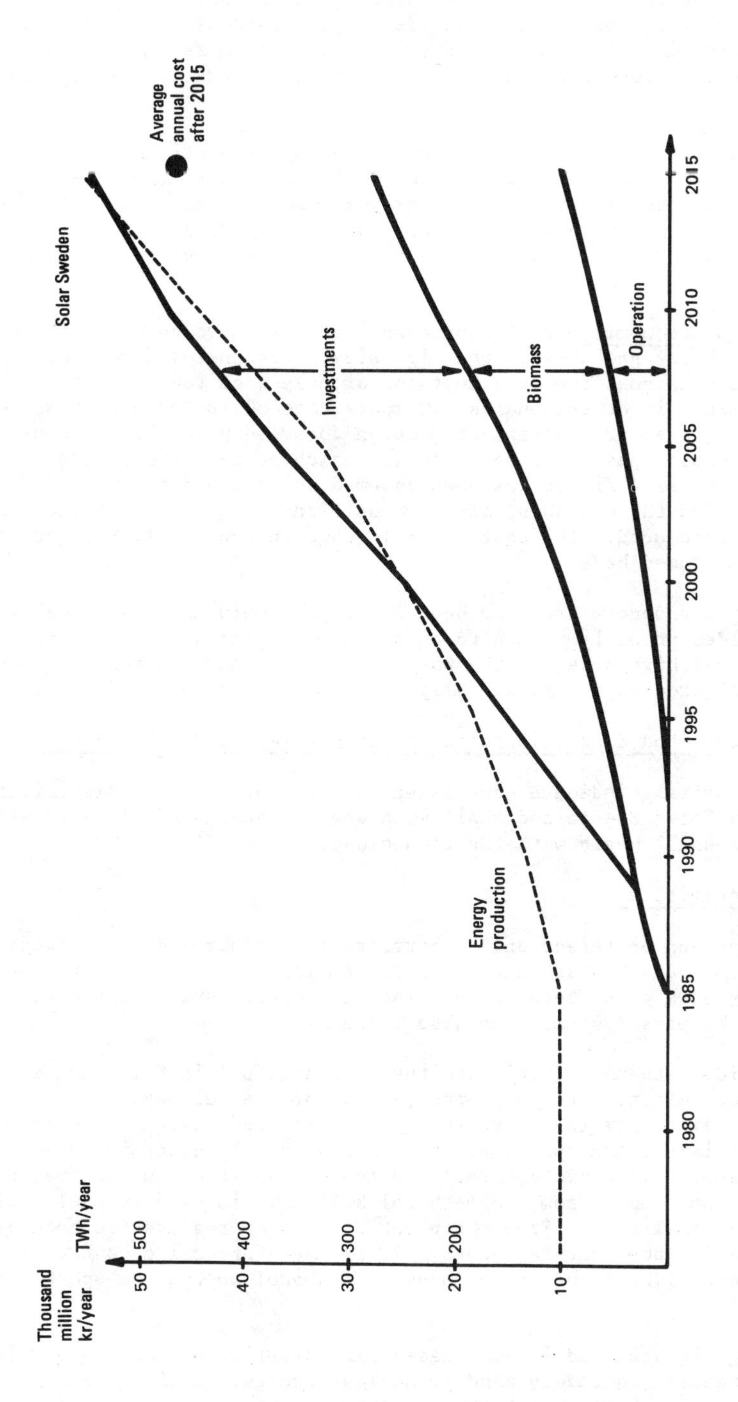

Figure 28. Energy contributions from and costs of Solar Sweden.

Another uncertainty concerns the efficiency of conversion from biomass to methanol or electricity and the possibilities of utilising the residual heat. The figures that we have used exceed what can be achieved today. But in several cases the assumptions we have made represent only moderate modifications of existing technology.

An important element in the energy system, where the uncertainties are fairly large is the fuel cells. The technology has been known for a long time and has also seen practical application, in the space programmes among others. Application in the energy system has been very intensively discussed but has not yet been realised. Major demonstration installations are under construction in the United States.

The uncertainty about economics is of course obvious. We have however, chosen cost statements which are probably on the high side. For energy forest plantations for instance, we have assumed the same cost as for pulp wood today, which is 100% higher than assessments in other studies. Compare note 30 to Table 22. We have not assumed any advantages from serial production for wind power but have used the estimated cost for prototypes. For solar cells, which represent over 40% of the total investments, the same figure has been assumed as that which the DOE has adopted as its goal for the mid-80's, although opinion now seems to be that these costs can be further reduced. The costs of Solar Sweden could well turn out to be lower than we have assumed here.

The environmental consequences are also uncertain. The main uncertainties with regard to Solar Sweden probably relate to energy forest plantations. There is, however, reason to set high hopes on the possibility that current research projects in this respect will produce environmentally acceptable solutions (12, 34).

Similarities and dissimilarities between Nuclear and Solar Sweden

We shall here very briefly indicate some essential similarities and dissimilarities between Nuclear and Solar Sweden and shall with several qualitatively very different dimensions. We shall begin with the technology.

Techological Characteristics

The distribution between different energy carriers in Nuclear and Solar Sweden is shown in Figure 29. Even this schematic representation shows large differences. The greatest differences exist between the district heating systems, the electricity system and the biomass system. See also Table 24.

On the transport side methanol constitutes the dominant fuel in Solar Sweden. A small share of direct electricity is planned, e.g. for the railways. In the uranium alternative either hydrogen or methanol is produced as fuel. In both cases fuel cells are used in the transport sector. These generate electricity and the vehicles are driven with electric motors. The two cases differ on the fuel side in that hydrogen is split off from the methanol before it is used in fuel cells, but this difference is slight . From the point of view of realisation both alternatives are similar for the next few years. It is therefore quite compatible with both main alternatives to begin introducing methanol in the transport sector now.

Heating of buildings is achieved in both cases with district heating in the larger cities. Electric heating is widely used in Nuclear Sweden. In Solar Sweden electric heating is used chiefly as back-up system in connection with solar heat installations. Common to both alternatives is the extensive use of district heating. The main difference is the temperature range in which heat is delivered

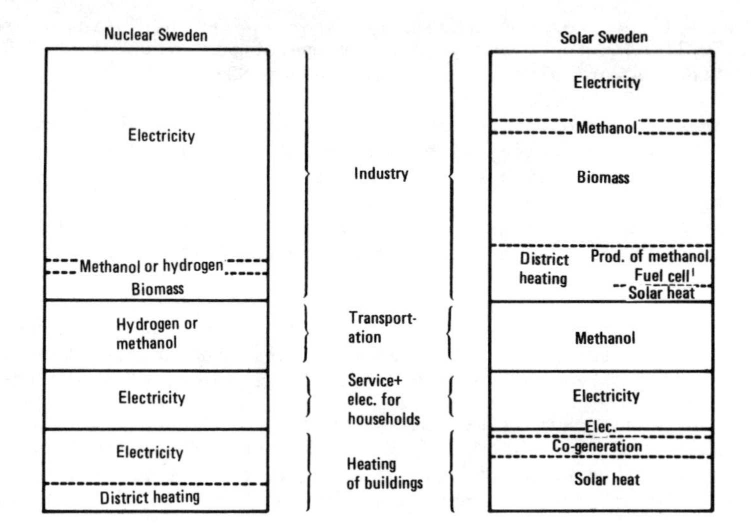

Figure 29. End-use energy carriers in Solar and Nuclear Sweden.

Table 24. Comparison of energy systems in Nuclear and Solar Sweden.

	Nuclear Sweden	Solar Sweden
District heating	Main urban areas (about 25)	Main urban areas (about 25) and all other large urban areas (about 100) as well as local systems for solar heat, fuel cells etc.
Technical demands	Somewhat lower water temperature (70–100°)	Lower water temperature (40–60° C).
Electricity supply system	Dominant. Small share of combined production (heat plus electricity). Direct space electric heating (and accumulating systems)	Complementary. Large share of production close to the consumer. Large share of combined production (heat plus electricity)
Technical demands	Greater need of peak power than today	Greater need of peak power than today
Solid fuels	Not used	For processing industries power and heating stations methanol production
Transportation	Hydrogen distribution (or methanol)	Methanol distribution

In the solar case heat comes from seasonal stores. On the basis of present technology these will need to be designed for temperatures around $40-60^{\circ}$ C. The district heating network of today is designed for $80-120^{\circ}$ C.

In Nuclear Sweden the production of process heat is done with electricity, in Solar Sweden through combustion of biomass. In the latter case the proportion of waste heat is higher, partly because of co-generation of heat and power, partly because of the extensive use of fuel cells in connection with the combined production of electricity and heat for industry. A significant difference between the two alternatives is thus how industrial process heat is produced.

One dimension of the technical system is the size of the units. Nuclear Sweden contains a small number of very large installations. These are furthermore intimately interconnected both through the fuel cycle and through the electricity supply system. Solar Sweden contains a very large number of small installations.

One dimension of the technical system is the size of the units. Nuclear Sweden contains a small number of very large installations. These are furthermore intimately interconnected both through the fuel cycle and through the electricity supply system. Solar Sweden contains a very large number of small installations.

Resources Requirements

The total costs for Nuclear and Solar Sweden are the same (cf. Tables 18 and 23). On the basis of the assumptions made about costs and introduction rates, however, Nuclear Sweden will require somewhat greater resources than Solar Sweden every year during the introductory phase except for the last years. This is partly due to the longer lead times for nuclear power. A certain amount of energy at a given moment requires a greater investment in Nuclear Sweden, which means that a larger share of the energy must be paid for in advance. There are costs for both Nuclear and Solar Sweden which have not been included but which are probably of the same order and therefore do not affect the overall picture. This applies to hydro power and the energy contribution from bark and lyes, and also to the distribution of energy (apart from investment in a developed electricity supply system). There is probably no decisive cost difference between operation and maintenance of the electricity supply system plus distribution of hydrogen in Nuclear Sweden and the distribution costs for biomass and methanol in Solar Sweden.

As we concluded earlier, there are substantial uncertainties associated with the future costs. For Nuclear Sweden we have used the present costs where they are known. For technology which is not available today we have made use of published figures (cf. Table 17). For Solar Sweden, because several of the techniques involved are at a lower stage of development, we have by and large used the costs of the prototypes. The cost of large-scale production may well be considerably lower.

If we calculate an average energy cost, it will be about 16 ore/kWh for both Nuclear and Solar Sweden, using 10% rate of interest and a 25-year write-off period for the investments at 1977 money values. Everything points to future average energy costs being considerably higher than today's. The total cost for energy today amounts to 20-25 thousand million kr per year. As we approach the year 2015, the cost of both the energy systems investigated will be about 50 thousand million kr per year. All the alternatives to oil are estimated to involve increases in energy costs.

Both alternatives are capital-intensive (i.e. the main part of the cost must be paid before energy is produced). If one postpones the introduction of capital-intensive technologies until they become profitable in terms of business economics (against the background of interest, depreciation rules, price of oil and so on) the availability of resources can become a serious problem. One must then pay for both the increasing oil import and for investments in the new energy technologies. The expected higher oil prices should therefore lead to earlier investments in technologies to replace oil. We have to seize the opportunity while the oil price is still relatively low. This is illustrated in Figure 30.

We shall now go over to a discussion of energy in the national economy. Can we afford to replace cheap oil? The answer is of course yes, but the question is whether production in general can simultaneously increase so much that the assumed doubling of production of commodities and services can really be achieved.

Energy in the National Economy

We shall restrict our study to the resource requirement expressed in the number of working hours on the supply side in Solar Sweden and Nuclear Sweden respectively, since the cost in ore/kWh is too dependent on future wage levels, institutional factors in the capital market, the form of enterprise, etc. The resources of a society ultimately consist of its citizens - money is merely an administrative tool.

Solar Sweden is no do-it-yourself society but both Solar Sweden and Nuclear Sweden require increased efforts in energy production. (Photo: Tiofoto).

By means of economic statistics we have estimated the total number of working hours for the different elements of Solar Sweden (35), as shown in Figure 31. Solar Sweden will require about 970 million working hours in 2015 for its energy system. If energy consumption is subsequently stabilised at the level of 570 TWh/year, 800 million working hours per year will be required on average for operation and replacement investments. In comparison we may mention that the whole of Sweden's energy supply today requires an estimated 350 million working hours per year. The diagram also shows the resource requirement for other energy, chiefly oil, during the period 1975-2015. We have here assumed that 1 m^3 of oil represents a constant number of working hours during the whole of this period.

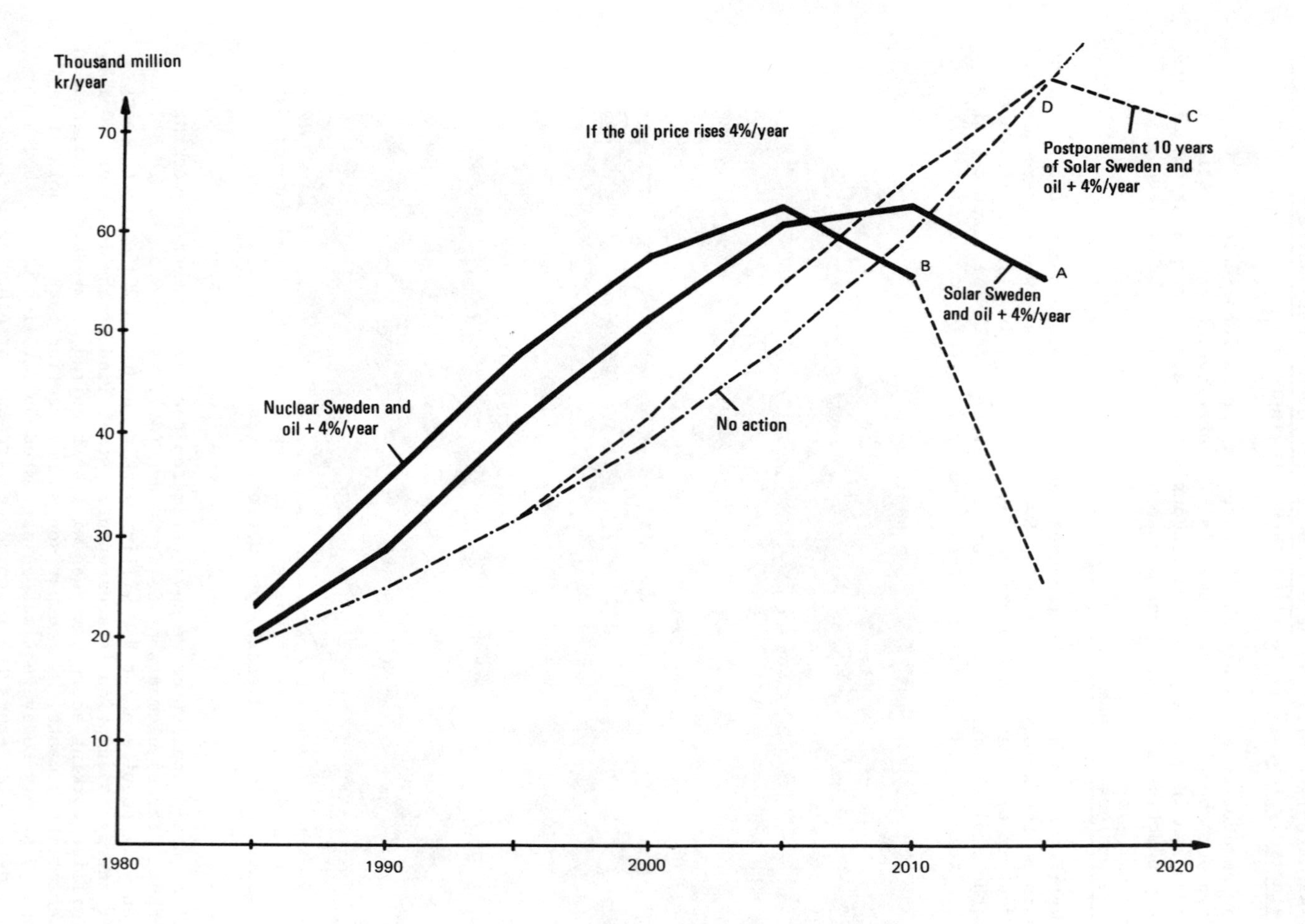

Figure 30. Annual costs of our total energy supply if the oil price rises by 4%/year and Solar Sweden (A) or Nuclear Sweden (B), respectively, is introduced as outlined in this chapter. The consequences if the introduction of Solar Sweden is postponed for 10 years (C) or if the present energy system is maintained (D).

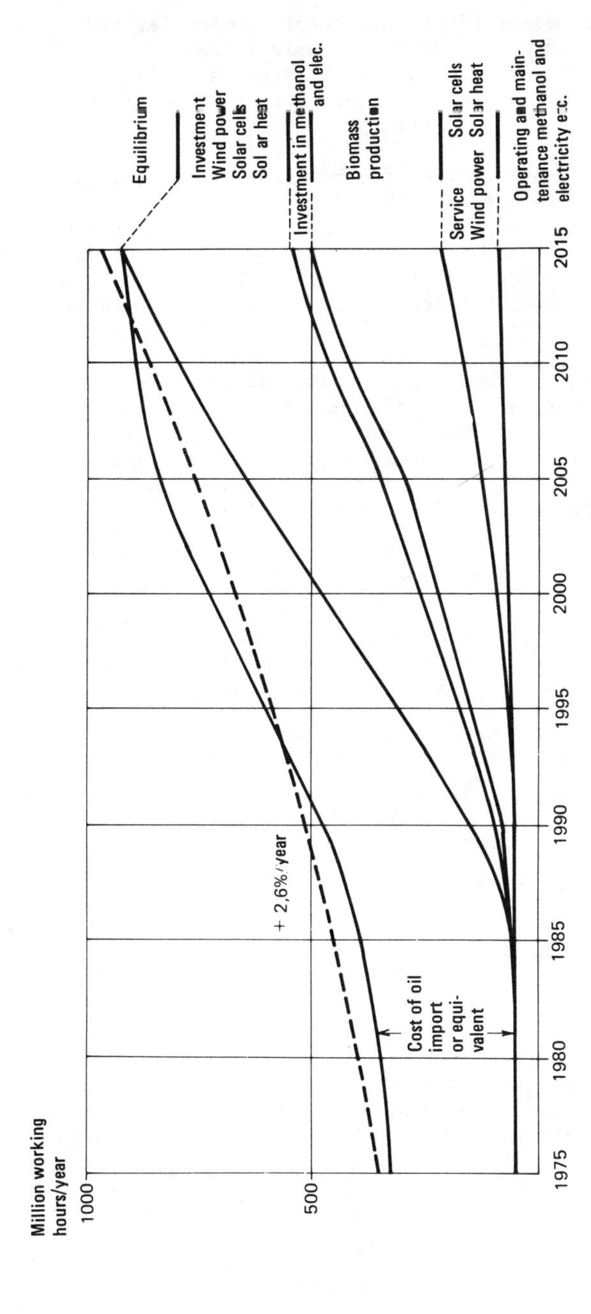

Figure 31. Resource requirement of the energy sector in Solar Sweden expressed in million working hours per year.

The development of a new energy system together with complementary energy during the period of expansion will demand an increase of manpower in the energy sector of, on average, 2.6% per year (from 350 to 970 million working hours per year).

Does this square with the doubling of production of commodities and services by the year 2015 which we assumed at the outset? Could it be the case that the labour required to provide the energy will be so large that the remaining labour force will be insufficient to achieve such a production target? The answer is th that it depends on the growth of productivity.

In order to achieve twice the production of commodities and services in 2015 (compared with 1975) a productivity increase of 1.75% per year will be required if the total labour effort remains unchanged. As the energy sector will absorb an increasing amount of labour, productivity in the remaining sectors of the economy must increase by somewhat more, namely by about 2%. Of these annual 2% about 0.25% (or about one eighth) will thus have to be diverted to the new energy system.

But is an annual productivity increase of 2% possible? We subdivide the enconomy into production of commodities, production of energy and services. For the energy sector we assume, as before, an unaltered labour productivity. We assume that labour productivity in the service sector will increase by 1% per year. We can then calculate the labour productivity in commodity production that will be requir- ed for the whole equation to balance. That will be 5% per year. The situation is shown in Table 25.

Table 25. National economy 1975 and 2015.

Labour effort for production of	1975 million working hours/year	Labour productivity %/year	2015 Production volume relative to 1975 %	2015 Million working hours/year
Commodities	2450	+5	+100	700
Energy	350	+0	+170	970
Services	3300	+1		4430
Transportation			+100	
Other			+100	
Housing			+ 40	
	6100			6100

The figures may be compared with the historical development.

Within the commodity sector labour productivity increased during the years 1960-1975 by 6.2-7.2%/year and is estimated (36) to increase by 5.4-5.7%/year during 1975-80. In the service sector it increased by 1.9-3.2% during the years 1960-1975 and is estimate to increase by 1.3-1.7%/year during the years 1975-1980.

There is every reason to believe that it will be possible to produce commodities more efficiently in future. An example of this is the use of mini and micro computers, industrial robots, etc. Biological production methods may expand considerably due to advances in enzyme and bacterial methods. Rising energy costs may possibly lead to a somewhat lower increase in labour productivity. The growth of productivity also depends on how quickly the production system adopts a more efficient structure. This ties in with regional policy and changes in industrial structure among other things. The increase in productivity is further dependent on the supply of capital, as a rise in productivity often involves investment in new machines.

According to the SCB[1] manpower forecasts the number of persons in the labour market is expected to increase (37). In that case the total number of working hours will also increase if a 40-hour working week is maintained. If the total volume of work remains the same in 2015 as in 1975, this will mean a shortening of working hours.

The conclusion to be drawn is that even if we in future acquire a much more expensive energy system, this does not necessarily mean that societal development in other respects will come to a halt.

The really important debate is about how society should make use of increasing production. Previously the increase in productivity of industry has been utilised to produce more commodities and to transfer manpower from industry to the service sector (medical services, child care, etc). There are naturally several needs competing for the increased productivity capacity. At present the energy situation (particularly with regard to oil) appears such that we must invest in a new energy system if we wish to avoid the risk of serious social problems in the future. An energy system such as Nuclear or Solar Sweden will, however, only claim a limited part of the increase in productivity.

If it should turn out that production does not increase because of structural problems etc, the conditions for introducing Nuclear and Solar Sweden may differ. A good deal of the energy techniques of Solar Sweden will not depend on specialists and will be geographically scattered. Investments in renewable energy technologies might possibly be a way of restricting unemployment. Most jobs could be handled by workers with experience of ordinary factory and building work.

Resources other than manpower. It is possible, of course, to make certain general calculations of other resources such as steel, cement, etc. There, too, the differences appear to be small. Solar Sweden would, of course, require considerably more glass (for solar collectors) that the present output, but there are good prospects for increasing glass production in the country.

Land requirements. Both alternatives differ greatly from the present energy system with regard to land requirements. We now import the major part of our energy but pay for it partly by utilising land in other sectors of production. Both of the future systems thus have in common a greater direct land requirement than the present energy system.

Sweden imports oil, which is partly paid for by the export of forest products. One hectare of land for timber production allows an import of 3-4 m^3 of oil. The yield calculated as m^3 of oil per hectare is thus lower for the ordinary forest industry than for energy forest plantations, which give 9 m^3 at a yield of 90 MWh/ha and year (35). Calculated in this way the oil import would correspond to an indirect land requirement in the order of 10 million hectares.

In the solar case large areas will be required, above all for the cultivation of biomass in various forms. To that will be added land for wind power, solar cells and solar heat. Potentially suitable land for biomass plantations is widely available. There are thus possibilities of siting the plantations with regard to other desiderata for land utilisation. Wind power is also flexible. The distance between towers can be varied. The number of towers will however be large, though still less than the number of power-line pylons in the nuclear case. Solar cells

(1) Central Bureau of Statistics

and solar heat will preferably be located adjacent to built-up areas and must therefore be included at an early stage in the planning of land utilisation in urban areas.

In the nuclear case a considerably smaller surface area will be required than in the solar case. Land will be needed for at least ten new nuclear power sites and for a greatly extended electricity distribution system. To that must be added land for mining, uranium extraction and installations for the other stages in the nuclear fuel cycle. The number of places suited for the siting of reactors is limited and will therefore not allow much freedom of choice. The same applies to electricity distribution - it will be necessary to construct power lines between the places of production and consumption.

The planning of land utilisation is one aspect of the institutional picture and we shall here touch upon a few more of the differences in this regard.

Structure of enterprises in the energy sector. Nuclear Sweden will require for its development and operation a series of co-ordinated decisions, which indicates the necessity of a single organization. The French model, with a wholly state-owned electricity supply sustem, a nuclear power industry with considerable state shareholding, and a wholly state-owned fuel cycle, is a natural form of organizational structure for Nuclear Sweden.

Solar Sweden will be based on another kind of technology which poses different requirements of co-ordination. The local adaptation between user and supply, but also the central adaptation between different supply components, poses demands which may perhaps be best satisfied by a mixed state/local authority/private structure.

Localisation. The reactors of Nuclear Sweden and installations for reprocessing and enrichment can probably only be dealt with in a national plan - the present veto right of local authorities must probably be abolished. Solar Sweden poses demands on land for biomass, wind power plants, solar heat generators. Overall, considerably greater areas of land - and thereby presumably many more landowners - will be affected by Solar Sweden than by Nuclear Sweden.

The property owners' choice of heating system must presumably be directed in both cases, but more radically in Solar Sweden. Bother Solar and Nuclear Sweden presuppose energy conservation and therewith reduced heat requirements for the heating of buildings. At the same time both Solar and Nuclear Sweden contain wide facilities for district heating. Energy planning is thus required in which investments are directed towards conservation or supply and not, without careful consideration, towards both in the same geographical area. This must affect the means for property owners to, for example, receive subsidies for energy-saving investments.

The capital market will differ. Nuclear Sweden presupposes that large amounts of capital will be channelled to a small number of large projects, while Solar Sweden presupposes also that large amounts of capital will be distributed but to households, companies, local authorities, etc.

The need for qualified manpower will vary. Typical for Nuclear Sweden will be the need for many highly specialised technicians, while Solar Sweden will be typified by great variety in the composition of manpower.

We can thus establish some hypotheses about the demands which the different solutions will pose on auxiliary activities. Table 26 should be see as a mental experiment which shows that the different solutions pose essentially different demands.

There are also other differences. But those already mentioned are sufficient to indicate that the different techniques pose fairly widely differing demands on the institutional framework. That is one of the reasons why the two systems will have difficulty in co-existing for any length of time.

One question is the range of different social forms comprised by Nuclear and Solar Sweden respectively. We naturally cannot answer that question, but we hope that the outlines of the two energy systems will widen the basis for a continued debate on the future society.

Table 26. Conceivable differences between Nuclear and Solar Sweden (40).

	Nuclear Sweden	Solar Sweden
Capital market	Large amounts of capital to be shared out among a small number of (a single?) recipient.	Large amounts of capital to be shared out among a great many recipients (households, property-owners, local authorities, firms).
Construction	Small number of large construction sites. Migratory specialists.	A great many small construction sites. Need for people to be capable of various tasks.
Division of responsibility centrally/locally	Local government veto right must presumably be abolished with regard to the energy sector's components. Stronger central powers vis-a-vis local authorities.	Certain rights and freedoms now enjoyed by individual households must presumably be transferred to local authorities. Strengthens local authorities vis-a-vis households etc.
Division of responsibility public/private	Interwoven interests administration/big firms.	Interwoven interests local authorities, residential areas local work-places.
Professional groups	Large demand for highly qualified specialists. Technocratic elite.	Large demand for broadly educated people with ability to adapt local energy sources to local conditions.
Central electricity system	Dominates its environment.	Dominated by its environment. Functions chiefly as standby and back-up system.

Knowledge of both Nuclear and Solar Sweden is inadequate. A definite choice today of either alternative involves taking risks which may have serious consequences. It therefore seems advisable to postpone the definite decisions for a number of years and to create freedom of action through positive measures

Variants of Nuclear Sweden and Solar Sweden

Nuclear Sweden and Solar Sweden, as they are presented here, are extreme alternatives and should primarily be viewed as examples.

The technical development on the solar energy side could make new technologies worthy of consideration (38). The possibilities of improving the efficiency in the use of energy have presumably been underestimated and the costs of different technical elements are uncertain. Taken together, this could very well lead to the realisation in 5 or 10 years' time that a renewable energy supply system should be designed in a different way than we have outlined here.

A uranium-based energy supply system is considerably less flexible, but it might be possible to postpone slightly the transition to breeder reactors. With the reactor types currently available - and with the time-perspectives applicable to the introduction of new reactor types - the range of choices is limited. If the time-perspective could be extended a couple of decades beyond 2015, which depends on the time limits imposed by the supply of oil, it is likely that new reactor types (e.g. thorium-based) could come to the fore.

For all of these reasons Nuclear and Solar Sweden, as they have been discussed here, should rather be seen as examples of two long-term and essentially different lines of advance. When we refer below to the nuclear and solar options it is these lines of advance to which we refer. We shall now proceed to discuss the forces which determine the line of advance.

5 Where are we going?

Introduction

In the last chapter we described two extreme lines of advance for domestic Swedish energy supply. We shall now discuss these alternatives from the viewpoint of present actuality. Both Nuclear and Solar Sweden (as well as combinations of the two) are attended by great uncertainties at present. We therefore regard it as reasonable to keep both options open, at least until the end of the 1980's.

Is that really a problem? Yes, if the present energy system is developing inherently towards one or the other alternative.

We noted in chapter 2 that the present energy supply system has largely been allowed to evolve on its own terms - the role of the state has been to smooth its path rather than to indicate its direction. There is nothing strange about the fact that organizations - whether they are private enterprises, state authorities or pressure groups - often act with great determination and wilfulness. The will to survive and evolve applies not only to individuals in their private lives but also to groups of individuals in a social context, e.g. business enterprises. And just as the will to evolve is a positive motive force for the individual, it is equally so for the enterprise or organization.

If the individual drive and private interests of the energy enterprises generally coincide with the politically desired development, there are no problems. But these arise if and when this coincidence ceases to exist. Then a power struggle begins, in which the side that is toughest, most tenacious and has the best knowledge of the conditions determining the future energy supply system, etc, will be in the best position to realise its aims and to present the other more or less with a fait accompli.

There are many examples of governments, in Sweden and other countries, which have time and time again, through a series of monor decisions over a long period, been faced in due course with a fait accompli where only one alternative remained. The aircraft system Viggen is presumably the most throoughly analysed case in Sweden (1).

The resistance to keeping several options open is understandable. Every decision about the future creates losers as well as winners. The industrial consequences of the solar option are quite different from those of the nuclear option. To keep both open involves uncertainty for <u>all</u> the affected parties. Organizational sociology has taught us how organizations (enterprises, authorities, trade unions etc) constantly endeavour to minimise the uncertainty of their conditions. Securit is probably more important than profit and the resistance to a policy which gives the state greater freedom of action will be considerable - from all directions (2).

We are furthermore dealing with long time-perspectives. In these such concepts as economy, profitability etc are extremely imprecise. Institutional conditions for example the rules of the capital market, land encroachment laws, and demands on the internal and external environment, acquire great practical significance for the success of one technology or the other. The time-perspective is furthermore so long that the technology itself is not unchangeable but can evolve and be adapted to various requirements. The technology of the solar option is considerably less developed than those parts of the nuclear option which are essential for energy production itself (the "front side" of the fuel cycle; see p. 46). At the same time the development of the solar option is in part quite different in character from that of the nuclear option.

Both options include technical possibilities which are difficult to assess. In this chapter we shall discuss the chances of evolution from the present situation of the solar and the uranium options. Our starting point is that of the general conditions applying to the introduction of new technology and the research and development that is already under way today. We shall subsequently indicate the plans and expectations of the energy enterprises and discuss the economic condition of competition for new technology. The chapter will be concluded with an imaginary example of stage-by-stage decision-making. The main thesis of the chapter is that we are already on the move and that the direction of movement is away from rather than towards a Solar Sweden.

The energy enterprises and the system of rules and regulations - a consideration of principles

There is said to be a light bulb which has a life of several decades and draws half as much power as the bulbs of today. Such a light bulb, if it can be produced at a reasonable price, would undoubtedly revolutionise the market for bulbs, shake up the light bulb industry and at the same time conserve energy. But irrespective of how revolutionary it is from a purely technical viewpoint it can hardly achieve success unless it has the same cap and is designed for the same voltage and voltage fluctuations as the old kind of light bulbs. It is very difficult to introduce a new technology which does not fit into the existing system.

The same applies to the domestic electricity supply. If someone wishes to install solar cells on the roof in order to generate energy for his own needs, this must be done in such a way that the cells fit in with the technical requirements of the refrigerator and with the delivery terms of the municipal electricity board, which in turn depend on the power producers. Energy-using appliances, energy producers and the manufacturers of equipment for energy producers are all parts of the same system. Changes at one point also affect other points.

The fact that many elements are interconnected means that they must be co-ordinated For this purpose there is a whole system of internal rules: quality requirements, product specifications, delivery terms, various kinds of contract, etc. These co-ordinating rules are mainly based on voluntary agreements between the parties. Perhaps the most important component of this system of rules is the principles governing tariffs and charges. The Swedish electricity sector has considerable freedom when it comes to establishing tariffs (see further the section "Tariffs and financing as conditions for competition").

We have now described an internal system of rules formulated by the parties themselves. All new technology must either fit into it or else whoever wants to introduce the technology must have the power to change the system of rules. The internal system of rules is, however, only a part of the overall system of rules governing the technical design of energy supply. We can identify two other parts.

The one that is easiest to describe is the energy policy system of rules. It consists of the rules formulated by government and parliament in order to direct and restrict the activities of the energy enterprises in various respects.

This system if rules aims to restrict a number of problems such as environmental damage, health risks, consequences for security policy, injustice towards weaker parties (price controls, regulations governing concessions, etc). There is here a whole complex of administrative laws (concerning environmental protection, work safety, military preparedness, building, electricity supply, water etc) which, taken together, affect the technical design of the energy supply system in various ways. It is not clear how strong this influence is, but a survey of government bills, laws, etc suggests that this part of the system of rules has a limited effect (20). It imposes certain restrictions and this renders certain technical systems more expensive or impossible (for instance the close siting of nuclear power plants or harnessing the hydro power of the Vindel river).

The regulations on energy conservation in existing buildings may constitute one significant exception. As they are framed at present, they favour a transition to electric heating rather than a district heating.

The third part of the system of rules is of major importance, however. We shall call it the general system of rules. It has only been established to a minor extent, specifically for the energy industry; furthermore it regulates the form of enterprise rather than the technology. It includes tax-law regulations (which determine which principles of bookkeeping are possible), company law, local government law (which regulates what local authorities can or cannot do), the control of the capital market by the Bank of Sweden, etc.

There are three forms of enterprise which are of particular concern in the energy sphere, namely state enterprises, private firms, and agencies or enterprises owned by local authorities. They operate under very different conditions. Perhaps the clearest example concerns the regulations which determine the financing conditions. A large part of the price paid for energy by the consumers consists of capital costs. The conditions for financing and "the cost of money" are thus of great importance. We shall return to these problems in the esction "Tariffs and financing as conditions for competition".

The overall picture of these three parts of the system of rules - the internal, the energy policy and the general - shows the first and the last as having the greatest importance for design of the energy system.

The energy policy system of rules determines within very broad limits what the energy enterprises can and cannot do, what initiatives they are prevented from taking, and so on. The general system of rules is of great importance to the economics of various technical systems, depending on how these fit into the conditions which apply to various forms of enterprise. The internal system of rules, finally, consists of the energy enterprises' own supplementation of the very wide-meshed net which is determined by the other two. This picture is far from complete. Above all it omits to deal with the way in which the different parts of the system of rules are based on and support each other.

The price levels set on electric power, for instance, are a result of, among other things, the yield requirements of the State Power Board (the energy policy system of rules), the conditions of financing (the general system of rules) and the electricity enterprises' own principles (the internal system of rules). If, for example, one wishes to alter the principle of price-setting in order to stimulate the introduction of a new technology or to stimulate savings, it will not only be necessary to transfer the internal rules, which at present are framed by the electricity industry, to the energy policy system of rules and to regulate them in law. In additions, the financial situation of the power enterprises, and thereby the capital market, will be affected. In the last resort the possibility of survival for certain enterprises may be called in question.

Our aim has not been to investigate the structure of this system of rules in detail. But we do wish to indicate that there are a number of rules, apart from what are traditionally called energy policy controls, which have great significance and that these rules are to a not inconsiderable extent framed by the energy enterprises themselves. The most important energy policy decision taken by the government and parliament has been to give the energy enterprises their large degree of independence. The ways in which this independence can be used has been illustrated by us in the case of nuclear power and co-generation (compare the section "Changing the source of energy - technology and organization" in chapter 2) (3). The lessons to be drawn from that are dealt with in the following section.

Technological change - the need for a driving force (entrepreneurship)

Both the nuclear and solar option will require a high degree of technical development. In this section we shall discuss the preconditions for technological change in order to analyse what institutional changes may be required for such a process of change to be amenable to control The discussion will be based principally on our studies of the evolution of the Swedish electricity system and the conflicts which have taken place in conjunction with it (see chapter 2).

There are several reasons why we started with the electricity sector. It is the only wholly domestic energy system and its development is therefore well documented while the development of the oil supply largely took place abroad. The electricity supply will play a key role in the future irrespective of how we go about replacing oil. It also has technical properties which lead to economic advantages through its interconnection into a large network, and with strict conditions that all suppliers and consumers must meet the same product requirements (voltage, frequency).

We shall begin by indicating some general conditions for technological change (3, 4).

There is an extensive literature on the driving force behind technical development and the introduction of new technology. According to one viewpoint it is the economic conditions which are the determining factor - a new technology is introduced if it is profitable (5). From another, more socio-psychological viewpoint it is clear that technological change often raises demands for new skills, while old forms of relations disappear and new social patterns emerge. Technological change is thus not only a question of commercial profitability but also of organizational stability and security for individuals.

Thus far the picture includes two factors and is relatively simple. Profitable technology is introduced though there is always a time-lag due to resistance to changes. In a wider perspective it becomes more complex. Profitability then depends on numerous factors which to some extent can be infludenced: the rules of the capital market, internal and external financing, the possibilities of altering and adapting prices and tariffs, demands on the external and internal environment. The conflict between nuclear power and co-generation is an outstanding example of this (3). New technology constitutes not only a threat to the social structure within an organization. It also affects the relations between different organizations, e.g. producers and consumers, the right to dispose of one's own or other people's land, etc. In a few cases it may be a question of breaking old cultural patterns or of influencing the class structure, as when the new Water Rights Act was pushed through against the agrarian interests in the Swedish countryside (6).

New technology here becomes a question not only of commercial profitability: from this viewpoint it is rather a matter of power and of the possibility of changing or making use of the whole (general, energy-political and internal) complex of regulations.

Every technical component - wehther it is a light bulb or a nuclear power station - must have an organizational basis. Somebody must produce somebody must use. New technology should preferably "fit in" with an already existing organization. Co-generation fitted in in Sweden because the Swedish towns distributed electricity and were in charge of district heating stations and networks. Coal in a gaseous or liquid form fitted in with the distribution system of the gas and oil companies, which makes it quite logical for these companies in due course to enter the coal industry.

If the new technology and the existing organizational pattern do not suit each other, somebody must assume the responsibility for altering the conditions. In the United States, for instance, nuclear power was established because the federal Atomic Energy Commission assumed responsibility for all stages of the nuclear fuel cycle apart from the light water reactor itself. In the same way the US government assumed responsibility for major disasters (the Price-Anderson Act). It is fairly certain that the introduction of nuclear power would have occurred much later if the electricity enterprises had been forced to take responsibility for the entire fuel cycle. The Swedish situation is like that of the United States. Nuclear power stations fitted more easily into the Swedish power industry, as it already included a strong central component with the State Power Board as the leading force.

There are certain differences between countries which have nuclear power on a large scale and those which do not. The countries which invested early in extensive power programmes for electricity production - Britain, France - have wholly nationalised electricity sectors, which thus act as a large buyer. The importance of large buyers for such complicated and capital-intensive technologies as nuclear power cannot be exaggerated.

Where then do solar heat, fuel cells, wind power, solar cells, energy forest plantations fit in? And what measures can one take in order for them to fit in?

This fitting-in is a question of the whole collection of regulations (energy policy, internal and general) which we discussed in the section "Energy enterprises and the systems of rules - a consideration of principles". What financial conditions should be valid for solar cells on the roofs of houses? Should the interconnection be made on the power producers' terms, on those of the electricity distributors or of house-owners? What tariffs should be adopted? The organizational basis for the various kinds of technology used in Solar Sweden remains to be established. To do so is necessary if freedom of action is to be achieved for this alternative.

Provided that there is an organizational basis, the conditions can also be created for a driving force - an entrepreneurship. We shall now discuss some conditions for the latter.

To introduce a new technology is a lengthy process. It is not a matter of a single decision but rather of many small ones and occasionally a few major ones. Often a whole series of problems has to be solved, legislation must be changed, a productive organization created and a market for the relevant technology developed. One of the most important tasks of the state is to create relatively stable conditions in order to provide time for an organizational basis for the new technology to be established. The analogy with the introduction of nuclear power is instructive. That was not only a matter of investments in research and development but also a question of legislation (Atomic Energy Act, Building Act, Radiological Safety Act, etc), of industrial policy (to develop a Swedish reactor production capability) and of creating and maintaining a market for electricity from nuclear power. This in turn required that competitors should be kept at bay (co-generation), that the use of electricity should be stimulated (tariff reductions) and that new markets for electricity should be opened up (electric heating).

When two or more technical alternatives confront each other a number of factors come into play. One of them is the question of who has influence over the market. The conditions for expansion of the electric power industry are determined by the overall level of electrification. That in its turn is affected by a number of factors, of which at least some can be influenced by the power industry. That applies, for instance, to the choice of insulating standards in houses, contracts between building contractors and local authorities concerning heating methods, and supply contracts with major industrial enterprises.

Another factor is the question of who controls the transport or distribution stage. For certain towns co-generation was a production alternative because they controlled both the distribution of electricity and the district heating stations and networks. At the same time the characteristics of heat and power stations were such that the towns could not be completely independent. They must be connected to the national electricity grid in order to gain access to standby power and to get an outlet for their surplus. As the State Power Board controlled the key point, i.e. conditions of entry into the main transmission grid, it was in a position to influence the economy of co-generation.

In general the companies which control several links in the chain (are vertically integrated) have an advantage over those which do not. Another example is the vertically integrated oil companies which hold a very strong position in the international oil trade (1).

Large companies also have other advantages. The large power producers have had a greater ability to finance their investments out of their own funds than, for example, local municipal energy producers.

Large companies are often able to set aside resources to bring influence to bear on those responsible for framing the regulations: the general, the energy policy or the internal ones.

That those who believe in and wish to advance a certain technology should also use the resources available to them in order to protect it is virtually a matter of course. One should neither be surprised at nor moralise about the fact that the State Power Board adapted its tariff levels in such a way that nuclear power was favoured and competing techniques became less profitable (8). On the other hand one may ask whether the power enterprises do not have to be deprived of parts of the freedom which they enjoy at present - if the government wishes to introduce a new technology which is contrary to what the power industry regards as natural and suitable. We have noted that, if two or more technologies are competing with each other, then that technology has an advantage which is represented by an entrepreneur with influence over the market and control of the distribution stage or who forms part of a large (possibly vertically integrated) enterprise. The nuclear option presently enjoys such an advantage over the solar option. There is an imbalance between the alternatives. This imbalance has the effect of causing the energy system to develop in the direction of Nuclear Sweden.

Research, development, plans and expectations

In this section we shall look at the distribution of research and development (R&D) investments in the energy sector and the plans advanced by the energy enterprises. We believe that this will illustrate which forces are active and their strengths.

Table 27 and Figure 32 present publicly financed R&D in six countries. They show that research and development in the world at large is chiefly directed towards fissil fuels and nuclear power. Much R&D is also financed directly by the energy enterprises. Table 27 shows an estimate of the distribution between public and private research.

Private research is of about the same overall magnitude as public research but has a different centre of gravity. The earlier a technology lies in the process of research and development - i.e. the greater the economic risks are - the larger is the share which is publicly financed.

Table 27. Publicly financed research in six countries (USA, Japan, West Germany, United Kingdom, France, Canada) and Sweden in 1976 (9).

	Six countries Thousand million kr	%	Sweden %
Oil and gas	650	4	1
Coal	1800	14	2
Nuclear power	8150	43	35
(thermal reactors)	(1650)		
(+ breeder reactors)	(3600)		
(fuel cycle)	(2100)		
Fusion energy	2050	11	7
Renewable energy sources	700	4	16
Conversion, storage, transportation of energy sources	450	2	4
Energy use and conservation	750	4	24
Other	4150	22	11
Total	18700	100	100

The above figures are naturally rough and partly conceal the fact that investment in renewable energy sources has increased very rapidly since the oil crisis of 1973-74. The establishment of the Energy Research and Development Commission in Sweden shows that the state has judged the basis of knowledge to be inadequate with regard, for instance, to renewable energy sources. The research grants have also been considerably increased.

Nevertheless the conventional energy sources, coal and nuclear power, still have an almost overwhelming superiority when it comes to the accumulation of knowledge - many times greater amounts than those now spent, e.g., on renewable energy sources have for many years been spent on the conventional ones. The superiority which the nuclear option has had and still has in R&D now contributes to the imbalance which

Table 28. Estimate of distribution of energy research and development between public and company financing in six leading industrial countries. All figures relate 1976 except for France where they relate to 1974 (9).

	Production						Consumption	Total %
	Short term		Medium term		Long term	Varying term		
	Oil and gas	Thermal reactors, nuclear fuel cycle	Coal	Breeder reactors	Fusion energy	Renewable energy forms	Industry Transport Housing comfort	
Public financing	15	65	85	85	90	50	15	55
Company financing	85	35	15	15	10	50	85	45
Total %	100	100	100	100	100	100	100	100

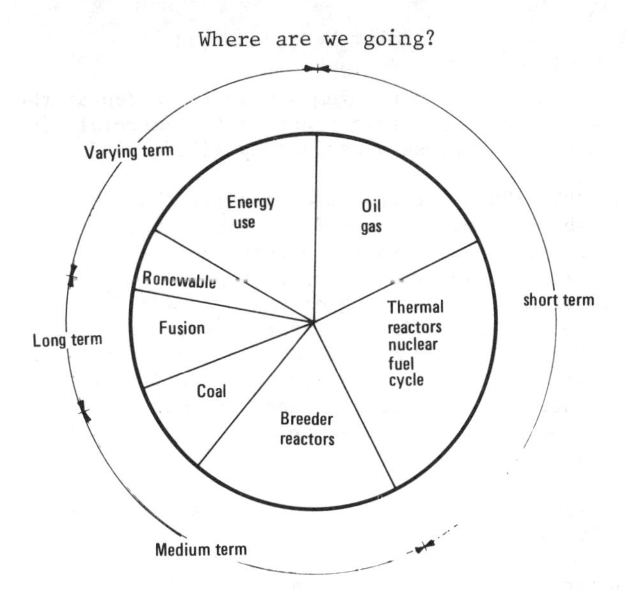

Figure 32. Estimate of distribution of energy R&D in six leading industrial countries (cf. Table 27). Total volume of R&D is about 26 thousand millions per year (9).

exists between the nuclear and solar options.

The international oil companies which are at present responsible for a large part of the energy supply are now beginning to orientate themselves towards other forms of energy as well (10). The oil companies are now rapidly developing into energy companies. Thus 24 oil companies, for example, control about 44% of the coal reserves for which mining concessions exist in the United States. Five oil companies control over 60% of the uranium mining capacity in the United States and 12 oil companies control 51% of the United States' uranium reserves (11). Four of the oil companies have concentrated their R&D investments in solar energy to photocells.

The major part of their R&D investments relate to the conventional energy sources shown above. The World Energy Conference, which is closely connected with the energy enterprises, expresses a virtually unshakeable faith in conventional energy sources - oil, gas, nuclear power, coal shale up the year 2020 (cf. Figure 12).

In the case of oil Sweden will be largely dependent on the global planning of the oil companies (7). It is only the electricity sector which has really been planned - by, among others, the planning committee of CDL. The Swedish energy planners have for the last 10-15 years generally envisaged a development in which nuclear power (including breeder reactors) would provide an ever larger share of the energy supply. In the government's budget for 1970 (12), for example, the Minister by and large supported the estimates of AB Atomenergi quoted there:

"In order to strengthen its position and develop its capacity, manufacturing industry regards it as necessary to make fast reactors commercially available together with fuel and other components for them, simultaneously with deliveries for thermal systems. The power industry in Sweden wished, in the same way as its equivalents in the main industrial countries, to obtain the economic advantages of a mixed system with a gradually increasing proportion of fast breeder reactors, by which it will become possible to exploit economically the plutonium output from the operation of the thermal reactors. The plutonium supply in Sweden at the beginning of the 1980's will permit at least one fast reactor of commercial size - about 1000 MWe - to be supplied with ready-for-service fuel by 1982".

The CDL study of 1972 (13), for instance, assumed that the total energy consumptic by 1990 would be about 75 Mtoe (about 870 TWh), of which electrical energy would constitute about 250 TWh. This would in turn require 24 nuclear power stations by 1990 together with an expansion of nuclear power and heating stations in the three largest cities by the beginning of the 80's and an additional one or two units during the rest of the 80's. The total installed nuclear power capacity by 1990 would then be about 25,000 MW, and by the turn of the century upwards of 60,000 MW (corresponding to about 350-400 TWh) could be installed. No reactors other than light water reactors were planned for during the 80's, but it was believed that gas-cooled high temperature and breeder reactors might become commercially viable during the 80's.

Electricity consumption would be distributed in the following manner by 1990 according to Table 29.

Table 29. Forecasts from CDL 1972 (13) and 1977 (14) for electricity consumption in 1990.

Users	1972 forecast for 1990 TWh	1977 forecast for 1990 (CDL-A) TWh
Industry	110	66
Services	50	26
Households	28	15
Electric heating	27	18
Summer cottages	4	3.5
Total including losses	250	153

It was estimated in the forecast of 1972 that the share of electrically heated single-family houses (as a percentage of the total number) would grow from about 30% in 1975 to about 75% in 1990. The CDL study is about six years old and we think that it gives a fairly representative impression of the thinking at that stage.

Of greater significance, however, is the fact that the Swedish nuclear power industry has been built up to about the level referred to in the study. According to its own statements ASEA-Atom needs a new order for a reactor every second year in order to maintain its capability and a new order every year in order to make ends meet in purely business terms (15). One order per year represents an increas of electricity production from nuclear power by about 60 TWh in ten years.

The electricity forecasts for 1985 and 1990 have been lowered considerably since 1972 and thus also the need for additional capacity for electricity production. This constitutes a direct threat to the Swedish nuclear power industry. A series of measures has therefore been taken by the power industry but also by the Federation of Swedish Industries among others. Among these measures are direct demands for the continued expansion of nuclear power and the extension of the nuclear fuel cycle within the country, demands for uranium mining to assist the export of reactors, propaganda against district heating and in favour of electric heating i single-family houses, and proposals for a change-over from the present energy taxation to a value-added tax. The latter would reduce the costs to industry for electricity but would increase those of households. As industry is more sensitive to price levels, this would presumably lead to an absolute increase in electricity consumption.

The Swedish power industry and a large part of Swedish industry in general has set its sights on a continued expansion of nuclear power. A very recent proposal as to how Sweden could phase out oil through nuclear power is based on similar ideas (Figure 33).

Up to now there has been no great interest in coal in Sweden. But if research should achieve such results that an increased use of coal becomes compatible with Swedish environmental requirements (disregarding the carbon dioxide problem), then the processing industry and the district heating and power and heating stations, for example, might conceivably become users of coal. And if the desire of the power industry to develop nuclear power were restricted, it is possible that demands for Swedish coal imports might also be raised in that quarter. We therefore cannot exclude the possibility that forces may later emerge which could pull Sweden into a new dependence on coal.

Taking into account how the energy enterprises of today think, act and argue, the only likely conclusion is that we are in the first place moving towards Nuclear Sweden, in the second place possibly towards an increased importation of coal. The direction of the movement is in any case away from rather than towards Solar Sweden.

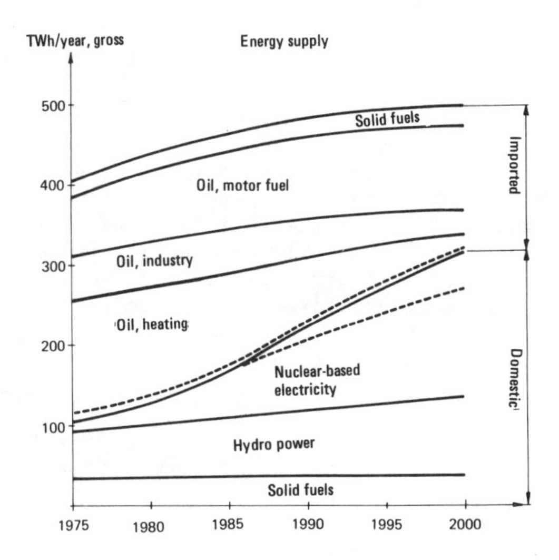

Figure 33. ASEA-Atom's proposal for expansion of nuclear power during the period 1985-2000 (16).

Tariffs and financing as conditions for competition

In the earlier sections we have discussed some conditions for the two energy futures, Solar Sweden and Nuclear Sweden.

To initiate the process which will make it possible in 10 or 15 years to assess with certainty and to choose one of the alternatives (or a combination of them) is not just a matter of research and development, however. Nor just of driving forces in the shape of organizations and enterprises. It is also to a large extent a matter of the economic conditions for the various technologies. The economics of the solar cells are determined not only by their technical properties, but also by the question as to whether production processes can be developed and the advantages of mass production can be obtained. Investment in production processes is only attractive, however, if the likelihood of profitable sales of the product is fairly high. This in turn depends on how well the production fits into the existing framework of regulations. We shall devote some attention to these questions.

Many solar techniques must be connected to the existing energy system, particularly to the electricity and district heating networks. The future of solar technologies is thus dependent on the conditions for connection provided by the established energy enterprises. But the effects of a measure are not always exactly those that are intended. We illustrate the problems by looking at the development within a related sphere: that of the water supply.

Translation from the Swedish daily "Dagens Nyheter"

New water rate penalises savers

Dagens Nyheter, Smaland correspondent

Alvesta, Friday

The water rates in Alvesta have been designed in such a way as to directly encourage wastage. All subscribers are forced to pay for a minimum consumption of 100 m^3, an amount which the majority of pensioners never use. There are cases of single persons who pay 25 kr for their actual consumption but who are charged a punitive extra rate of 275 kr.

– – –

The extension of the water supply (17) during the 1970's was based on forecasts from the 1960's. These did not take into account any assessment of price elasticity. The large investments had to be financed, however, so the water tariffs were raised (producer's measure 1). The result was that consumption did not increase in step with the forecasts (consumer's countermeasure 1). The producers were then left with existing investments which had to be paid; the water tariff was therefore raised even higher (producer's countermeasure 1). The consumers' reaction did not fail to materialise and consumption fell even further (consumer's countermeasure 2). The producer then changed his strategy. He altered the construction of the tariff so that the consumers paid a higher fixed rate - irrespective of how much water they consumed (producer's countermeasure 2). The consumer is now in a position where his water rate bill is unaffected by how much he consumes and the producer has succeeded in making the forecast self-fulfilling. The result is shown in the box above.

This example has its equivalents in the electricity and district heating sectors. The fundamental problem is that the units of the system are very large, extremely capital-demanding, and have to be planned long in advance. The possibilities for the energy enterprises to adapt themselves to other situations than those foreseen when the expansion took place are therefore slim. The action taken by the water and electricity companies in raising the tariff when people consumed less than expected is logical from the viewpoint of the need of the companies to repay the investments that they have already made.

It may seem natural to have tariffs for electricity and district heating which stimulate savings. It would then involve a redistribution so that the fixed rate is lowered and the variable one raised. Such tariffs have been introduced in some other countries. But they face great resistance from the electricity enterprises (20) as they later may have difficulty in covering the investments that they have already made.

The problems connected with energy saving sometimes also occur with new energy sources.

The interest in solar-heated houses is growing rapidly, e.g. in the United States. Houses which are 100% solar-heated are very expensive today, however (because of the storage problems), and it may be expedient to have say, 70% sun and 30% of something else, e.g. electricity. In Colorado the electricity tariffs were for historical reasons set at a level where this combination became economically advantageous compared to houses which were entirely heated by electricity (19). If a large number of consumers install such systems, however, that would have serious consequences for the electric utility. The latter would then be forced to maintain the capacity to supply a lot of energy for a short time in the winter while being unable to supply any energy at all for heating at other times. The need for installed capacity would therefore remain high, while the quantity of energy supplied would be low. The costs of the electric utility would by and large remain the same. But with the existing tariff the revenue would then have become too small in relation to costs. Consequently the electric utilities requested permission to change their tariffs so that a higher rate would be charged for installed capacity and a lower one for energy. The result - whether intended or not - was that the partially solar-heated house now became more expensive than the entirely electrically heated one.

Both examples (water rates and solar hear) illustrate the same fundamental problem. The design and planned extension of the present electricity system tends to increase the importance of large, capital-intensive power installations. The producers are best served by regular and assured sales. A development of that kind can bring about a tariff structure which creates difficulties for the incorporation of new production alternatives.

Heat pumps have the characteristic, among others, that their efficiency is reduced when the difference between internal and external temperatures increases. Electrically-powered heat pumps are unfavourable to the electricity producers for the same reason as solar heat was in the above example. Certain German electricity enterprises therefore demand that electrically-powered heat pumps must not be used during the winter, but that heating must then be done, e.g., with oil or gas.

One can guess that wind power will face the same problem. If a local authority wishes to develop wind power it must also have access to standby power. But from the viewpoint of the standby power supplier/wind power means a reduction in the utilisation period, i.e. the same costs but lower revenues - unless the tariff is altered. If that is done, however, wind power may become unprofitable for the local authorities.

The above examples have concerned the electricity system, but the same kind of problem occurs with district heating. Even if it is quite feasible technically to combine district heating and solar heating this can lead to the same kind of problem as that described above in the case of solar heating (Colorado, USA). If the district heating network is moreover served by a co-generation plant the contradictions are exacerbated. The investments in the latter are often so large that the local authorities have a strong interest in maximal exploitation of the investments. The consumers are therefore induced to utilise district heating as much as possible rather than to save.

There are reasons, however, why problems of integration are particularly difficult for the electricity system. Electricity cannot be stored but must be produced at the same time as it is consumed. As the demand for electricity varies in time for different users, there are advantages in integrating many users in one system. In this way the electricity generating capacity can be utilised better and the costs can be lowered. An established producer can also drive out competing industries, etc. which attempt to produce their own electric current (24). The problems experienced by many local authorities when they have tried to establish co-generation are an additional illustration of this. The economics of electricity production improves if one connects to a larger grid. At the same time this has to be done on the conditions for interconnection which exist and which have been created by the established producers. One can only market energy which complies with established rules. This central determination of "product design" does not exist in the fuel sector for example. There various kinds of fuels can be marketed and also find buyers. Heavy oil, for instance, which was previously regarded as a residual product in the production of motor fuels in refineries, has found a use in large central heating plants, etc.

The examples have demonstrated the extent to which the costs of fixed assets - the investments - can determine what solutions are arrived at. A major part of the price which consumers pay for energy in fact consists of capital costs. The financing conditions - how one obtains investment funds and what they "cost" - are thus of great importance. Investments are now financed in part externally, through the capital market, in part internally, through the revenues of the enterprises.

The construction of Aero-generators may present a new alternative energy source.

From an energy policy viewpoint the capital market can be said to consist of four parts, namely the financing of housing, the financing of power producers, the local government loans market, and the rest. Different energy techniques can be fitted in various ways into these four parts.

The situation with regard to a single-family house is illuminating. All installations in the house are financed through the housing loan system (long mortgage loan plus final loan, possibly also state home-ownership loans). The systems for supplying energy to the house are, on the other hand, mainly financed by other means

We shall look at three alternatives.

The individual oil-fired boiler is financed through the housing loan system. It therefore has more favourable loan regulations than the alternative of oil-fired district heating. In the latter case the linking of the house to the connection point (serving one block) can also be financed through the housing loan system, while culverts and district heating plants are financed in other ways. These financing possibilities were previously very bad - local authorities had to resort to the local government loans market - but now it is possible in principle to finance these items through debenture loans (SPINTAB and others). The problem is that permits for the issue of debenture loans are not granted at a sufficiently fast rate at present. If the house is heated by direct electric space heating, the major cost of the heating system is the power production (21). The power industry has precedence in the capital market and the investments incur lower capital costs than those of district heating.

Table shows what a wind power plant may cost under various financing conditions.

Table 30. Costs of projected wind power generator with different forms of financing (22).

Energy cost, ore/kWh at median wind	"Housing loans" 6% interest (year 1) 30 years amortisation, annual instalment 0.073	"Power industry" 10% interest, 25 years amortisation, annual instalment 0.11	"Bank loans" 14% interest, 75% marginal tax, amortisation with equal amounts 10 years	"Inflation-adjusted loans" 2% interest, 10% index, amortisation with equal amounts 20 years
5 m/s	15	22	26	15
7m/s	9	13	15	9

Another example is electricity produced by back pressure technique in an industrial enterprise compared to electricity produced with nuclear power. If the nuclear power station is financed through the state budget, the State Power Board will pay 10% interest on the capital. The nuclear power station at Forsmark is financed through debenture loans with a repayment period of 23 years at 10%, which gives an annual instalment of 11.5%. An industrial back pressure plant, on the other hand, may be financed with a loan at 15% interest and a repayment period of 10 years (23), which would for example correspond to an annual instalment of 20%. The capital cost per invested krona will then be twice as high for back pressure as for nuclear power.

At the same time it is not so easy to equalise the conditions of competition. Debenture loans require securities and as security one must have real estate. The back pressure plant which is an integral part of an industrial process is not regarded as real estate in the sense of the securities law, however, and consequently it cannot be financed in the same way as nuclear power. But the incentive system for energy savings can be altered in order to influence the conditions of competition.

The capital market at present favours a division between energy producer and energy user. In the interest of more efficient energy utilisation the exact opposite would be desirable.

But the picture is even more complex than that. Investments are partly financed internally by the energy enterprises - during the 1960's the State Power Board earned enough money to enable it to finance almost 70% of its expansion itself.

The greater the proportion that is financed internally, the less dependent are the energy enterprises on the priorities on the capital market.

A high degree of self-financing is possible because the tariffs of the energy enterprises are set according to depreciation on the replacement value and not according to the capital costs. Energy enterprises with a large proportion of old plants still in operation have a clear advantage here over enterprises with a cmall proportion of old plants. Nationally organized energy enterprises - whether they are state-owned or private - also have a clear advantage over locally organized ones. The State Power Board can finance the construction of power plants in a local authority area through writing off investments made elsewhere - a possibility which is never open to a local authority.

The different technical components in the solar and nuclear options enter the electricity supply system in different places. We have previously noted that there are big differences in financing depending on whether it is undertaken by households, distributors, the proprietors of transmission line grids or producers of bulk power.

We showed above (Table 30) how the economics of wind power varies considerably depending on which form of financing one selects. Solar cells in buildings are another borderline case. One variant is to finance them in accordance with the Housing Loans Ordinance, another is through the distributor. The former type of financing results in about half the capital cost and thereby half the electricity price compared with that of the latter.

The financing conditions thus constitute a significant aspect in the assessment made by anyone considering investment in electricity production (or conservation measures). Another aspect is what it would cost to buy the power from outside instead. The tariffs offered by distributors and/or producers of bulk power play a decisive role in this respect.

For wind power at the local authority level it will be advantageous if the contracts between the producer of bulk power and local authority provide for low costs of back-up power (i.e. when the wind is not blowing) and relatively high prices for surpluses (what the local producer transmits into the grid on a windy day).

Nuclear power plants have very high fixed costs and low variable costs. The producer of power is concerned about high utilisation and wants to have contracts which stimulate consumption in order to safeguard sales of the power produced.

There is a clash here between two possible principles of tariff fixing. Should the tariff be based on the conditions set by investments already made or on the conditions which favour the introduction of new technology? Producers of bulk power and distributors who have to meet demands for repayment of investments made design their tariffs primarily in accordance with the former demand. Such "cost-true" tariffs have a clearly conservatory effect - they favour the introduction of technology with the same cost structure as the existing one (e.g. large proportion of fixed costs and a large proportion of the total cost to the producer of power).

The conclusion is relatively clear. An electricity supply which is to stimulate development and the introduction of technology such as solar cells, fuel cells, wind power, co-generation etc poses other conditions on the rules governing tariffs, contracts, delivery requirements and quality demands than do light water reactors and breeder reactors. The rules demanded by the latter will automatically be disadvantageous to the former.

Who then will decide what form the contracts should take? The relations between subscribers, distributors and bulk power producers are only to a limited extent regulated by Swedish law. It is above all the revenue demands of the State Power Board and the self-financing requirements of the Local Government Act which are significant in this context.

What is decisive is therefore the practice worked out by distributors and bulk power producers, particularly the State Power Board, and the extent to which new techniques will come into conflict with this practice.

The present-day energy enterprises not only have sufficient resources themselves to develop or order new technology; they are also in such a strong position that they can bring considerable influence to bear on the competitive conditions for technology which does not fit into their own pattern.

Piecemeal decision-making - an intellectual experiment

A development towards a Nuclear Sweden - which need not have exactly the form outlined in chapter 4 for the year 2015 - may well result from a fairly common-place series of piecemeal decisions which may be quite undramatic at each individual stage. It could for example look like this.

1978 Direct electric space heating is accepted as the form of heating for all single-family houses. Vigorous conservation programme for existing housing. Reactors 11, 12 and 13 are constructed in order to preserve ASEA-Atom.

1979 Energy taxation is converted to value-added tax in order to make the export industry more competitive. Industry, which is more sensitive to prices than households, is allowed reduced electricity prices while those for households are raised. The net effect is to make the export industry more competitive and to produce greater electricity consumption.

1980's The economic difficulties of most local authorities make it very hard to expand district heating. Energy conservation also reduces the basis for expansion of district heating. This gradually reduces the prospects for rapid development of solar technologies on the heating side.

1981 Decision in principle on Swedish uranium mining in order to develop capacity on the shale handling side, to reduce the costs of importing uranium and to improve the trade balance through a certain export of uranium. Mining begins at the end of the 80's.
Continuing research on renewable energy sources. The price of coal rises more slowly than the price of oil and new coal firing technology makes it possible to burn coal without major environmental problems. Certain increase in the use of coal, e.g. in industry. Time is thereby gained for research on energy forests, and decisions on problems of land utilisation can be postponed.

1983 — Economic boom. Energy demand increases faster than expected, especially for electricity. Decision in principle to build another 2-3 reactors for electricity production during the 80's in order to maintain Swedish capability in the nuclear power industry.

Mid-80's — World trade boom leads to increased oil consumption all over the globe. This results in disruptions of production, oil price rises, and certain disruptions of supply. The economic boom gives way to a slump. Renewed interest in Swedish uranium enrichment. West Germany, for instance, can offer technology and perhaps finance in exchange for guarantees of uranium supplies. Domestic enrichment is thought to improve the balance of trade and to increase domestic capability.

End of 80's — Surplus capacity in electricity. It becomes unprofitable to introduce, e.g., cheap solar cells or wind power, especially as this would require large investments. (Bulk power producers who have a surplus have no motive to invest in wind power, and others have no access to capital). The solar heat technologies are now technically developed. The tariff structure makes them somewhat more expensive than electricity and, above all, applicable only in new housing. Existing single-family housing therefore converts to electric heating, as the district heating network which was to make solar heating competitive was never constructed in the areas of single-family housing.
The processing industry increasingly converts to coal. The forestry industry develops new methods of utilising whole trees and processes for utilisation of new fibre products. Competition for land for energy forest plantation increases.

Early 90's — International agreements on reprocessing. Under strict international control this is, after all, thought to provide better protection against the proliferation of nuclear weapons than a large number of central stores for spent fuel. Swedish decision in principle on a reprocessing plant for Scandinavia. Provides an opportunity for learning about the plutonium cycle and for recycling of uranium and plutonium. This makes it possible to export more uranium, which improves the balance of trade. A number of older heat and power stations (Vasteras, Linkoping, Orebro) are replaced by nuclear heating stations. Several additional nuclear condensing power stations are ordered.

Mid-90's — Surplus of electricity. The price is reduced in order to utilise fixed investments. New industrial processes are developed for electricity. Industry gradually replaces coal by electricity. Synthetic fuels are imported for the transport sector.
Cooperative projects with West Germany are discussed with regard to breeder reactors in exchange for Swedish uranium export. Provides the chance to learn a possible technology of the future and to preserve freedom of action. Fuel cells and solar cells are cheap but there is no market as there is a surplus of electricity.

Late 90's — Adequate supply of plutonium but simultaneous shortage of uranium calls for investigation of a breeder reactor programme for the twentyfirst century.

This course of events, which should only be see as an example, may not lead to precisely the Nuclear Sweden which we outlined in chapter 4 but to a nuclear power system which may in time become extensive. It does not preclude certain renewable techniques such as solar-heated houses in certain areas of new housing, but these play a subordinate role. It illustrates how energy alternatives develop (if a double metaphor is permitted) as a combination of zip-fastener and express train. A series of "natural" decisions on limited aspects are taken continuously and the same zip-fastener mechanism which led to the light water technology during the 1960's gradually hooks us into a system in which oil is replaced by coal (or natural gas) and nuclear power. That is, in fact, the direction in which the energy enterprises - oil enterprises, coal producers, mining industry, electrical industry, electricity producers etc - are attempting to move with the force of an express train which has gathered speed, and that is the direction in which governments and authorities of all major industrial states are endeavouring to go. The reason is quite simply that these techniques "fit into" the existing tariff system, organizational patterns and power structure.

6 The Transitional Period - on Energy Supply in the 80's

Introduction

The time-perspective of energy policy is long. It took 50 years for the share of coal to rise from 25 to 50%, it took 25 years for oil to increase from 20 to 70%, it took at least 20 years to establish nuclear power.

Even if we made the decision today, a new energy system would not play a dominant role until around the turn of the century. As far as we can judge today, there are only two conceivable principles for a lasting energy supply based on domestic sources, namely the nuclear and solar alternatives.

Both alternatives are chiefly situated around and beyond the turn of the century - but at the same time the decisions which we take during the next decade will be determinant for the future. That is when the distribution systems and the industrial structure will be created, the capability to order and use plants will be built up, and legislation and organizational forms will be adapted.

A Swedish energy policy must, however, also consider other than the domestic alternatives. One should not, for instance, a priori presume that coal cannot becom an increasingly large element in energy supply - not because of deliberate politic decisions but by way of the same kind of non-decisions as have led up to the present dependence on oil.

We noted in chapter 2, and expanded on it somewhat in chapter 5, that the role of the state in energy policy has hitherto been to smooth the way and to define limit rather than to indicate the direction of advance. In this and the subsequent chapter we shall now discuss how this tradition can be replaced by an active and guiding role which appears to be a necessity if the talk of freedom of action is to be really meaningful. The starting points will then be

<u>that</u> Sweden's energy supply in the long term is to be based on domestic energy sources

<u>that</u> the choice then lies between the solar and nuclear options and possibly combinations of these

<u>that</u> both options today contain considerable uncertainties

<u>that</u> energy policy is directed towards not closing the door on any of the options

<u>that</u> the nuclear option today has a considerable lead over the solar option

<u>that</u> an open-door policy therefore involves making the solar option a comparable alternative in practice

<u>that</u> at the same time the energy supply of the 80's and 90's must be safeguarded.

Freedom of action must not be confused with freedom from acting. To keep open the decision of future energy supply systems requires a number of wise and decisive choices already today.

To keep the door open will be difficult for several reasons. There is always strong opposition against <u>not</u> taking an immediate decision from the group which has most to lose, namely <u>all</u> those who have an interest in the nuclear option. The latter option has today a large lead over the solar. Not to take any further decisions therefore automatically favours the nuclear option. The endeavour to preserve the Swedish nuclear power industry, for example, can lead to an expansion of nuclear power and electrification motivated by industrial (rather than energy) policy which can make it significantly more difficult to introduce a renewable energy system at a later stage.

High demands will thus be placed on the <u>transitional system</u> which will be a link in time between an oil-dominated and an either solar- or nuclear-dominated energy system. We believe that, on the basis of the argument in chapter 5, the demands can be summarised as follows:

<u>Buy time</u> for the longer-term decisions! This involves on the one hand economising better on energy, on the other replacing oil by other energy sources in order thereby to limit oil consumption. In the long term it is a matter of allocating a subordinate role to oil in Swedish energy supply.

<u>Stimulate technical development</u>! This is chiefly a matter of R&D for the renewable energy sources. Reasonable guarantees are also demanded in order to ensure a market for successful development projects. If, for example, the conditions for integrating solar technologies are unsuitably planned, nobody will wish to invest in them.

<u>Create flexibility</u>! It must be made easy in future to integrate solar heat, as also electric heat, for domestic heating without large additional costs. The construction of flexible systems of utilisation is in part a purely technical question, but is also connected with the principles for the conservation of energy.

<u>Check on the organisational and institutional structure</u>! It should, by and large, be neutral in relation to the long-term alternatives. It is neither possible nor desirable that all private enterprises and authorities should be neutral. People and organizations often work better if they are convinced that their particular solution is better than a competing one. But we ought to construct an institutional system in which the forces pulling us in the direction of the nuclear option are balanced by forces which pull us towards the solar option. It is important to ensure that the conflicts and contradictions of interest which inevitably exist really balance one another.

In the remainder of this chapter we shall discuss the likely consequences of the listed demands for energy sources and energy utilisation during the 1980's. In chapter 7 we shall subsequently discuss the organization of energy policy more thoroughly.

Energy sources during a transitional period

To what extent do we have to complement oil with other energy sources during a transitional period while waiting for the nuclear or solar alternative to take over completely? The answer to this question depends on how quickly energy utilisation increases, how quickly the nuclear or solar option can expand, how much one can conserve, but above all on the assessment of how the oil supply will develop.

In order to establish a few starting points we shall refer back to the calculations in chapter 4. If we deduct the contributions made by Solar and Nuclear Sweden respectively from the total supply level, we obtain the energy which must be provided from other sources (cf. Figure 34). As the energy contributions of Nuclear and Solar Sweden grow at roughly the same rate we have only shown one curve (for Solar Sweden). Today the remaining energy supply consists almost entirely of oil.

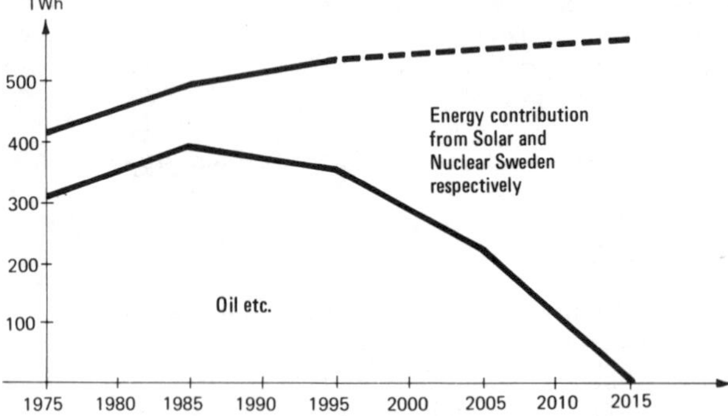

Figure 34. Remaining requirement of energy supply after deducting the energy contributions of Solar and Nuclear Sweden from the total energy requirement. Total supply level according to the forecast of the National Industrial Board (alternative A) (1) up to 1995 with a slower increase thereafter.

The diagram shows that if we do not take any steps to replace oil, the consumption of oil will reach a peak in the mid/late 80's and subsequently decline. It is possible that this pace in "the march away from the oil society" is sufficiently fast, i.e. that a decisive turn towards Nuclear and/or Solar Sweden would be sufficient. But we may nevertheless ask how it could be speeded up.

We know something of the role that conservation of energy can play. In the forecast of the National Industrial Board (1) there is an alternative forecast (alternative B) which is based on larger contributions to the energy conservation programme. That would involve a saving of about 35-40 TWh per year by 1995 in relation to the main forecast (alternative A). Further savings are possible, but only with increasing costs (cf. Figure 35). A well designed energy conservation programme has clear advantages over an energy supply programme. The environmental effects are insignificant and there are no uncertainties corresponding to those relating to supply. On the other hand there are other elements of uncertainty: knowledge about the savings to be obtained for a certain input of resources is very inadequate.

By what can we replace oil on the supply side? In the Swedish case there are only three or four kinds of energy which could produce a major volume during the 80's: nuclear power, coal, natural gas and peat.

The realisation of the present nuclear power programme (10-13 reactors) can be

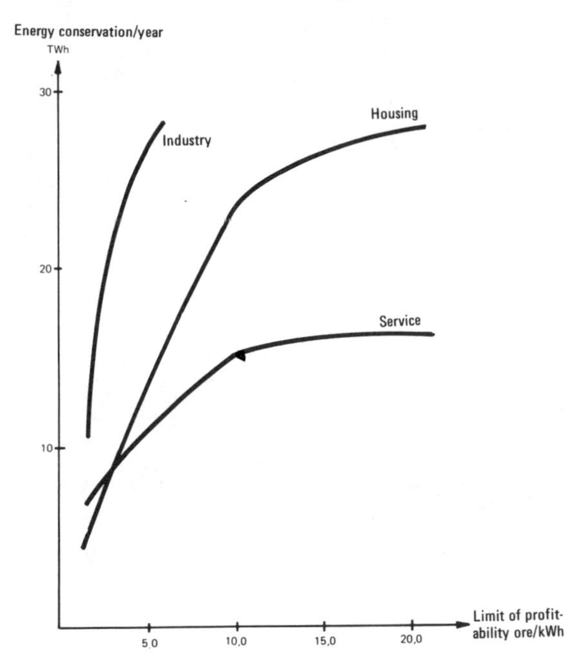

Figure 35. Potential savings from the present situation as a function of cost (2). The diagram shows the energy saving which is profitable at a specific energy price ore/kWh. This is a net energy price, i.e. the price at the utility, e.g. heat from an oil-fired boiler.

looked upon as a way of limiting oil consumption. These reactors should be discussed separately from a nuclear option as they belong to a shorter time-perspective in which major blockages already occur. Nuclear Sweden, on the other hand, is and ought to be a future alternative which should be maintained as a choice for some time yet.

We have discussed coal in chapter 3 and drawn the conclusion that coal could hardly be an alternative on a large scale and in the longer term for Sweden. The problems relating to environmental effects and trade policy are too great. This should not prevent us, however, from discussing coal on a smaller scale, a few million tons a year, equivalent to perhaps 20-50 TWh. One should also note the possibility that industry, for example, might prefer to import coal rather than use other fuels or electricity. Import control may then become necessary.

Natural gas has also been discussed in chapter 3. Here, too, the conclusion is that this energy source cannot play a major role in the long term. The admission ticket in the form of investments in infrastructure is expensive and the long-term uncertainties of delivery are a reason for limiting the programme. In practice only Scania and possibly Western Sweden would probably be eligible. There is no gas distribution network in either Nuclear or Solar Sweden, and investments in such a network should then take this into account. It is of course feasibly to produce gas from biomass and distribute it in pipe-lines, but the conversion losses would probably be large. In practice it may be possible to import one or a few tens of TWh a year in the form of natural gas. Swedish importation of methanol produced from natural gas, on the other hand, would be worth considering as an alternative.

Peat is a domestic source of great potential. The total peat reserves in Sweden are estimated at an equivalent of 3000-4000 Mtoe (about 30,000-45,000 TWh (3)), of which maybe 10% could utilised. These 300 Mtoe could during a period of 20-30 years provide, on average, 10-15 Mtoe a year (110-170 TWh), which is undoubtedly a large contribution. Peat has a large volume per energy unit and therefore high transportation costs. The handling during production and utilisation of peat also requires considerable investments. The processing industry, co-generation plants and heating plants are possible peat users. It would be conceivable to guarantee a certain peat market through legislation (compare the compulsory heating by wood-fires in government buildings during the 30's). One use could be for methanol production. Peat handling could provide valuable experience for the utilisation of biomass (in Solar Sweden). An established organization for peat handling could also facilitate rapid expansion during a serious energy supply crisis. The selection of peat-mosses to be taken into production must be done with careful consideration forlocal ecological factors.

Figure 36 summarises the measures that have been discussed above to replace oil. They could mean that oil consumption after the mid-80's would fall by about 40% during the subsequent ten-year period. By 1995 it could be reduced to half the 1978 level.

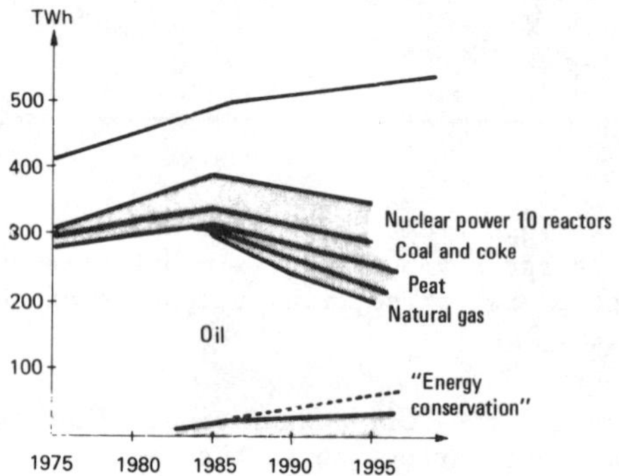

Figure 36. Reduction of oil dependence through conservation of energy and utilisation of other kinds of energy, apart from investment in Nuclear and Solar Sweden (cf. Figure 34).

The real possibilities of importing oil, and likewise coal and gas, are difficult to assess. It is part of an active state oil policy to increase the security of supply and to spread the risks. Such a policy could comprise agreements with producer countries (e.g. barter trade agreements such as the delivery of Swedish industrial systems in exchange for oil contracts) and participation in the work of prospecting and extraction.

Eaergy utilisation during the transitional period

In chapter 4 we analysed the similarities and dissimilarities between Nuclear and Solar Sweden. We would now like to take those observations as the starting point for a discussion of the kind of energy utilisation that will be possible in a transitional system. It must be designed in such a way that energy supply at a later stage can evolve either towards Solar or Nuclear Sweden (or a combination of them). To build in this possibility of choice probably involves additional cost compared to making the decisions now. The extra cost will undoubtedly be said to be unnecessary, wasteful and irresponsible in terms of the national economy by those who have already made a choice. Such a transitional system will, on the other hand, be cheaper than if an early decision, whether directed towards Nuclear or Solar Sweden, should later have to be changed for one reason or another.

We shall study this problem on the basis of a rough division of energy utilisation categories, as follows:

- transportation
- industrial processing heat
- heating of buildings
- electricity

In this we have departed to some extent from the traditional description, which usually proceeds from industry, transportation, services and housing. Heating of buildings also includes the heating of industrial buildings, while electricity includes all uses of electricity. Industrial processing heat can be used as a (waste) heat source for the heating of buildings. There are thus overlaps and interconnections between the spheres of use.

Transportation

The transport sector is at present based on oil, in Solar Sweden on methanol-operated fuel cells and in Nuclear Sweden on hydrogen- or methanol-operated fuel cells. Distribution and the fuel systems in vehicles are therefore basically the same, while there are differences with regard to the production of the fuel. This difference is not critical.

Methanol as a fuel can be introduced at an early stage and can be produced during a transitional period from oil, coal, natural gas or peat.

The production of liquid fuels from biomass or by means of nuclear power involves considerable conversion losses. The transport sector may therefore pose great demands on the energy supply. These demands would be reduced by more energy - efficient transportation, which is therefore of importance whether we choose Solar or Nuclear Sweden.

It is a striking fact in itself that the forecasts assume that thefuel consumption of private cars will increase 13 times more than that of public transportation (14.4 TWh a year as against 1.1 TWh) up to 1990 (4). That creates a dilemma for economy measures in the transport sector. How can (structural) procedures to reduce the need for cars be weighed against and combined with procedures to improve the efficiency of individual cars? Neither aim is particularly easy, as the transport sector is rigidly integrated in relation to both.

The transport system is integrated into community planning, the labour market and the daily lives of private individuals. It is necessary to view the transport sector as a single system in which different means of travel can both replace and complement each other. The car can sometimes be replaced by public transportation, sometimes by bicycles or walking, sometimes not. The need for rapid - and long-distance - movements results from the greater division of labour and specialisation. The rapidly growing lorry traffic over long distances is at least partly a consequence of changes in the internal production planning of companies, the effort to keep down storage costs, etc. Structural changes take a long time and have far-reaching effects.

The automobile and oil companies operate in international markets, which places particularly strict limits on what a single small country can do. Can we have a policy of our own for technical design or is that regarded as a technical obstacle to trade? Engine design and fuel supply are furthermore adapted to each other and the international oil companies dominate the latter.

An active energy policy for the transport sector involves the introduction of methanol and must,as far as we can understand, imply a partial disengagement from the international system as organized at present and a change in the rules governing the daily life of the car buyer/user. This may be difficult and will take time.

Industrial Processing Heat

The energy needs of industry (apart from the heating of buildings) consist largely of processing heat (see chapter 1). The methods of supplying this heat differ markedly between Solar and Nuclear Sweden, as shown in Table 31.

Table 31. Energy supply for industry (TWh/year)

	Today (1974)	In 2015 Solar Sweden	Nuclear Sweden
Electricity	38	71	218[4]
Oil	71		
Methanol		10	10[3]
Biomass (direct use)		86	
Other fuels[1]	51	36	36
District heating[2]		44	0
Solar heating		17	0
	160	264	264

1. incl. bark and lyes
2. waste heat from fuel cells and methanol production
3. alternatively hydrogen
4. not corrected for eventually of using electricity more efficiently than fuel in processing heat applications (cf. Figure 19).

Solar Sweden also contains combined plants to a much greater extent (processing heat up to 500° C from methanol production, fuel cells etc). Nuclear Sweden is far more streamlined from the technical and organizational aspects.

Present-day industry uses heavy oil to a large extent. This oil is used in large central heating boilers, particularly in the processing industry, and also in district heating plants, power and heating plants and condensing plants. The consumption is large enough to be able to carry quite considerable fixed costs in the form of distribution and handling systems.

It is conceivable that heavy oil could be replaced by gas, coal, peat or biomass. This is already done in part today - the utilisation of chips and bark has increase in the paper and pulp industry. In Finland certain heating stations and power stations are fired with peat, as also in Ireland. During the transitional period one should endeavour to

- reduce the specific fuel consumption
- replace heavy oil by other fuels
- prepare for the future use of biomass

The processing industry has a number of interesting features:

- the number of installations is small, around 200 at present using a total of 120 TWh per year

- these can to a greater extent than other energy users produce electricity themselves (back pressure) or utilise waste heat or blast-furnace gas in combined projects

- the energy costs of the processing industry are high. It is therefore more price-sensitive than other energy users.

- the processing industry has gradually become more efficient in the use of energy and there are big differences between old and new installations. The same applies to employment, environmental pollution and working environments. The rate of structural change in the various branches is of great importance to future energy utilisation.

- several of the branches constitute the backbone of Swedish inustry in general and that of certain regions in particular (Norrland, Bergslagen).

If solid fuels are to be introduced in Sweden the processing industry, district heating plants and co-generation plants will be the obvious users. If heavy oil is to be replaced by solid fuels, their introduction will be eased if the market for the new fuel is constantly expanding. Added to this will be the possibilities for the processing industry to supply waste heat and to participate in energy combines.

The use of energy and choice of fuel by the processing industry can be influenced in various ways - e.g. through the price of energy, costs and availability of capital, or through concession ruels. The fact that relatively few installations are involved makes it realistic to have a licensing system for energy use in existing installations, similar to that now provided for under §136 a of the Building Act in the case of new installations. In that way the planning of industrial and local government energy production and use can be co-ordinated.

Heating of Buildings

The sector concerned with heating of buildings differs considerably as between Solar Sweden and Nuclear Sweden and also diverges from the present form of heating, as indicated in Table 32.

Table 32. Different kinds of energy for heating of buildings (energy quantities supplied). Percentages (rounded off).

	Today	Solar Sweden	Nuclear Sweden
District heating	10	20	30
Heating of individual houses or blocks			
Oil	80	0	0
Solar heating	0	70	0
Electricity	10	10	70
	100	100	100

By block heating we mean central heating plants for one/several blocks. With solar heating there is an incentive to interconnect about a hundred flats into a common heat store, whereby the whole heating requirement could be met from solar heat.

The housing sector presents considerable factors of inertia. Houses stand for decades and the distribution systems also have a very long technical life-span. During a long period we have to both save energy overall and change distribution systems. But saving, supply systems and new production alternatives are interrelated and affect each other. We shall take two examples.

Nuclear Sweden contains nuclear thermal reactors in certain towns. These are expensive and the costs are largely unaffected by size. Intensive application of conservation measures can reduce the heat load and thus impair the economy. The same applies to power and heating plants.

Waste heat utilisation faces the same problem. Nykoping could, for example, be heated through a waste heat pipeline from the Oxelosund ironworks - the fixed costs for this would, however, be so high that the waste heat alternative is only economical if the heat load is not too restricted. In Nykoping funds for energy conservation should therefore not be spent primarily on reducing energy consumption in houses but on extending and connecting the district heating network to a waste heat pipeline from the ironworks at Oxelosund.

Our conclusion is that the heating of buildings should be directed towards the demands posed by Solar Sweden, i.e. water- (or air-) borne heating systems in private houses as well as a combination (in the urban areas) of district heating and block heating. Such an arrangement could also be utilised in Nuclear Sweden. The waterborne systems hould be adapted to lower water temperatures than today. Energy conservation and the extension of the energy supply must be analysed and vary carefully co-ordinated. Co-ordination of that kind does, however, require that the present legislation should be changed. We shall return to that in Chapter 7.

This may be compared to what will be the natural course if the nuclear option is chosen from the start. Then very large investments will be devoted partly to saving energy in single-family housing, partly to extending the distribution network for electricity so that electric heating can be used in every house. In all new single-family housing direct electric space heating will be introduced.

It is not particularly easy to compare the economics of the various alternatives. An example from Stockholm (5) shows that the cost difference at present between introducing district heating and electric heating may be small in existing single-family houses. It may increase, and to the disadvantage of district heating, if large resources are simultaneously devoted to energy conservation. To convert to a waterborne heating system after building for direct electric space heating is both expensive and difficult.

The State Power Board's studies (6) show that in a large nuclear power system (moving towards Nuclear Sweden) an accumulating system (i.e. with thermal storage) at the consumer's premises would be profitable because it makes possible a distribution of the electricity demand between day-time and night-time.

There do not then appear to be any decisive economic reasons opposing waterborne district heating and block heating in the long term.

Electricity

A transitional solution for electricity supply is without doubt the most difficult problem. The design of the electricity system differs so markedly as between the nuclear and solar alternatives. In the former the electricity system is a completely dominant part of the overall energy supply, in the latter it is in absolute terms smaller and has a partly different structure. Table 33 shows this (energy supplied).

Against the hundred or so reactors of the nuclear alternative one must place perhaps a million solar cells in connection with individual houses, a few thousand fuel cells for combined production of electricity and heat, a few thousand wind power generators and a few dozen power and heating plants of the solar alternative. A number of hydro power stations are common to the two alternatives.

Table 33. Electricity production in TWh/year.

	Today (7) 1976	Solar Sweden	Nuclear Sweden
Hydro power	54	65	65
Nuclear power	15		400
Power and heating plants (not nuclear)	7,5	13	
Industrial back pressure		6	
Fuel cells		24	
Wind power		30	
Solar cells		50	
Condensing power (oil)	7,5		
Total of electricity supplied	84	188	465

A comparison of the forecast of the National Industrial Board for 1995 with the energy consumption of Solar Sweden shows that the electricity level in Solar Sweden somewhat exceeds the forecast for 1995. We would therefore not need to restrict the overall electricity consumption level during the next decade. The task is rather to control how energy is consumed. Cases in which fuels or solar heat could be used later must be designed in such a way that this could happen. We have previously also discussed processing heat and heating the organizations and regulations which control electricity supply in such a way that they do not eliminate Solar Sweden.

The two alternatives do, however, have certain common features and these are fortunately sufficiently numerous to form the basis for the technical development of the 80's and 90's. The electricity system is in both cases interconnected into a network covering the whole country. Both alternatives need technical components, e.g for load distribution, and here pump storage plants, fuel cells and accumulating hot water systems on the users' premises could be common components.

We have discussed the conditions for introducing solar techniques into the electricity system in the section "Tariffs and financing as conditions for competition" in chapter 5. The conclusion drawn there is that the problem is also an institutional one - the conditions for financing and connection into the system must be suitably formulated. As this could in turn create difficulties for the electricity producers of today, a fundamental change in the economic and financial conditions of the electricity sector will presumably be required. The present freedom of the electricity producers to establish tariffs etc must be subordinated to the objectives of the energy policy and not, as at present, determine it.

7 Organization of Energy Policy

Introduction

We have previously in chapter 5 noted that powerful forces are pulling us towards a situation where the nuclear power option appears increasingly inevitable. If freedom of action is to obtain, something must be done. In chapter 6 we discussed which technical demands one ought to make on a transitional solution that would make it possible during the 80's to choose Nuclear Sweden or Solar Sweden.

We must now discuss how the necessary decisions can be taken. For that a partly new organizational structure will be required. It should be capable of stimulating technical development, creating the conditions for the use of products on a sufficiently large scale and taking the risk of failures. It must be sufficiently centralised to be able to remove rapidly any economic, legal and organizational obstacles and to give powerful support to new possibilities.

It is quite clear that neither "the state" nor a few authorities, organizations or enterprises will be capable of this: it must be a combination, a whole system, a broad process.

The best parallel to what is required is perhaps the electrification of Sweden. That was a process which for about 40 years, between 1890 and 1930, brought into being a number of institutions and laws to cope with conflicts between the demands of technology and the reality of that time. Both a driving force - the construction and power industry - and a strong political and social will to change were required. Electrification was as much a social as a technical issue.

Political will can never be legislated into being. But through legislation and other measures one can make it easier for a commitment to be pursued. Here, too, there are interesting lessons to draw from the period of electrification: the political will at the national level met with a response from the cooperative movement for electrification, above all in rural areas, which gave rise to a number of associations for the distribution of electricity.

In the rest of this chapter we shall discuss what form an organizational structure may take. We do so with hesitation, however, for two reasons. Firstly we must naturally query the pattern which applies today. This will worry and irritate many; we are probably all rather sensitive when our social relations are queried.

Electrification in Sweden 1890-1930 forced into being a number of new institutions and statutes in order that conflicts between the demands of the new technology and the demands of that period could be solved. Electrification became as much a social as a technical question. (Photo: Pressens Bild).

Secondly, despite the fact that a fair amount of research has been done, our knowledge about organizations in interaction ("multi-organizations") is rather inadequate.

But we shall try nevertheless.

Who can take the responsibility?

In chapter 5 we discussed the importance of there being a driving force - an entrepreneurship. Some person or persons must feel an overriding personal responsibility for introducing the renewable energy sources in particular. Only then is there a chance that all the existing obstacles will at least be identified, perhaps also overcome.

One conceivable force is the present-day electrical energy enterprises, among which the State Power Board has a dominant role. With their present background and activity they function by and large as entrepreneurs for the nuclear option. It is therefore highly doubtful whether they can also act decisively for the solar option. To us the electrical energy enterprises do not appear suitable for implementation of an energy policy directed towards freedom of action for two reasons.

The first is that this role presupposes coordination and adjustment between different interests. For that the present-day electrical energy enterprises largely lack the capacity, while there are other organizations which have experience in handling similar problems.

The other reason is that the present-day electrical supply enterprises derive their position and their role from a very "heavy" energy system (with long lead times, evident indivisibilities and large capital needs). The energy sources and energy carriers which are to be promoted have other characteristics and are furthermore competitors of those which the present-day electrical energy enterprises wish to develop.

An organizational basis could instead be sought in the local authorities, which are at present responsible for a large part of the coordination in the community, We can see good reasons for starting with them.

The amounts of energy which the present-day energy enterprises regard as "uninteresting" are often of low thermodynamic quality (of the waste heat variety), occur in small lots and are irregular or difficult to anticipate. We therefore need an organization wich also takes care of these when they appear as waste heat, (potential) back pressure, wind power, solar energy, or small hydroelectric plants in local streams and rivers.

Renewable energy sources are also space-consuming (2). They therefore require extended land planning, both in urban areas (for solar heating systems), in sparsely inhabited areas (wind power etc) and also for forest and agricultural land (energy plantation forests). Should one, for instance, wish to combine the heating of buildings and waste heat, the possibilities will depend, among other things, on where in the local authority area the installations are placed.

The nuclear option (2) also has land requirements, but they are more concentrated and relate to uranium mines, sites for power plants and land for power transmission lanes. Both the nuclear and solar option will pose greater demands on long-term land planning than the present energy system does. But they pose different demands and this should result in different kinds of planning and decisions. The land planning and decisions for the nuclear option must take place largely at national level. For the solar option it will to a considerable degree be a matter of local decisions.

There are local authorities which pursue ambitious energy plans. But there are also those which have shown a weak commitment to energy questions. It is by no means unproblematical to assign the chief responsibility both for introducing renewable energy sources and for administering energy conservation to the local authorities and their energy utilities.

The local authorities have been given an even larger role during the post-war period. To create a basis for the renewable energy sources within ten-twenty years will involve a very extensive evolution of local authority capabilities. But within 20 years, since the mid-50's, the local authorities' capabilities have increased greatly. There is no reason why this could not happen again.

The local authorities obviously cannot develop these capabilities themselves. A wider local responsibility can only grow up in interaction with consistently organized support - economic, legal etc - from the government and the state administration.

The organizational structure, the general features of which we discuss below, consists of:

- local government responsibility for supply, conservation and utilisation of "own" energy sources to the largest possible extent

- state responsibility for a strategy for phasing out the dependence on oil. This may include concessionary legislation to replace heavy oil by other fuels. Also active consumer policy and technical stimulation through government orders.

- financial survey of the energy sector with the aim of guaranteeing sufficient capital and distributing it effectively from the viewpoint of energy management. Also a survey of laws etc which regulate the tariff structure of the energy enterprises.

- reorganization of the electricity sector. Distribution could be made a purely local government responsibility and the authority of the State Power Board could be transferred to a newly created organization.

Local government energy planning

According to the fundamental approach which we have outlined the task of local government energy planning should be to take responsibility for all energy supply within the area of the local authority. By analogy with the principle of energy economy which was laid down in 1975 at national level, this responsibility should embrace not only the construction of distribution systems but also the balance between supply and conservation. The inhabitants and enterprises of the local government area should be stimulated to utilise solar energy, heat pumps, wind power, combined production of electricity and heat, utilisation of waste heat from industry, etc. The local authority should to the least possible extent "shift" the problems onto others and import energy across the local government boundaries in the form of fuels or electricity.

Legally such a role could most suitably be based on the present law on local government energy planning. In its present form, however, this is mainly a law on local government energy supply. The idea of conservation is there, but must be more clearly formulated.

Organizationally we believe that an attempt should be made to develop the local government energy utilities in that direction. They should through legislation be given the task of successively building up and offering the inhabitants of the area a combination of energy supply and conservation measures. The coordination tasks could include solar heating for individual houses or blocks and wind power generators, e.g. propertywise (if these turn out to be realistic alternatives).

Such an extension of their tasks will be neither easy nor natural for the present-day energy distributors. They have mostly, for quite understandable reasons, regarded conservation as a threat. But for that very reason it is necessary to try and get the distributors to become involved in conservation and to integrate it into their activities.

Such a development has partly occurred in the United States, above all with the gas distributors but also with the electricity producers (3). Their reason was that it was easier to restrict the need for further supplies through conservation measures than to produce these supplies. The conflicts over new power stations have been so intense and the development of demand so uncertain that the conservation option has on the whole had advantages. The energy enterprises have made credits available and offered the consumers packages of measures which the latter then repay through electricity bills etc.

Conservation should be organized in such a way that people can control it. A local government organization is thus better placed than a state authority. This is not only because the local authority is less distant from the citizens, but the Local Government Act provides considerably greater scope for the detailed control of local government activities than to the powers granted by the Swedish constitution to the government to control state authorities.

At the same time it is clear that wider powers and duties for the local authoriti also raise a demand for reforming, and in part widening, the state regulation. We shall return to this. But a fairly immediate consequence would be that the present system of energy conservation subsidy should be changed. Instead of bein a subsidy applied for by property owners for a specific house, it should be a subsidy/loan applied for by the local authority for a specific geographical area. *If* the local authority's plans show that resources should be applied to the savin of oil in individual propterties (and not for example to district heating or a waste heat pipeline), then the property owner could apply for a subsidy from the local authority. We refer to the example of Oxelosund/Nykoping.

We shall end with a brief reminder about the various levels of management. Local government energy management must begin with the individual *property* (or even fla within the property). The property owner is formally responsible for the energy management. Consumption depends on operation, maintenance and technical design. In the latter we include the heated space, insulation standard, location, the design of the heating systems, etc. These can in part be influenced, e.g. throug building standards and housing finance. But the possibilities for the local authorities to control the need for energy are today too limited.

The *block* constitutes the next level in energy planning. Groups of properties ca be planned so that they are supplied with heat and possibly with electricity from a central plant for each group. Town planners should take into account that properties in suitable large groups can be connected to a central heating plant. The latter can then be charged, for instance, with solar energy, surplus heat fro adjacent industry, a diesel generator or conventional district heating plant. Th should also be possible to carry out with large parts of the existing housing (an has already been done in some places). New housing areas should be planned in such a way that a possible future use of solar heat or solar cells is facilitated. One example of such measures would be to orientate roofs or wall surfaces in a suitable direction for solar irradiation.

The next level consits of *urban sectors* (or other large parts of the local authority area). There one can sometimes count on the possibility of utilising industrial waste energy for heating of buildings.

At the *local authority and inter-local authority* level the energy management planners must consider how much ought to be devoted to systems for energy production and transmission and to energy conservation measures respectively. Decisions must also be taken regarding "import" and "export" of energy across the local authorit boundaries.

Particularly *heavy-energy users* (nuclear power plants, major processing enterpris etc, perhaps around 100 in the whole country) should naturally be integrated into the local energy management system. But they are often so technically complex and so large that it is doubtful if they should be wholly subordinated to the local authority energy management. At present new installations are being tested by the government in accordance with paragraph 136 a of the Building Act. Such installations will in many cases concern more than one local authority (district heating from Stenungsund to Gothenburg, from Barseback to Lund, from Oxelosund to Nykoping). For this reason among others these should continue to be subject to central planning.

The interaction between local government energy plans and, e.g., the processing industry needs to be regulated. Today state authorities have certain powers to demand that an industry supplies waste heat to a local authority. On the other hand there is no way of forcing a local authority to construct a district heating network if waste heat, for example, should be available. In the same way as the state has been able to oblige local authorities to construct water supply and sewage systems it ought also to be possible to force local authorities to develop district heating, for instance, if that should be in the national interest. The state should presumably also be given greater powers to promote the formation of combines between industry and/or local authorities.

One problem (with a fundamental significance going far beyond the energy sphere) is the choice of organizational form for local government energy activity. At present this is mainly conducted either in the form of local government joint-stock companies or of local government departments. The joint-stock company form provides greater freedom of manoeuvre with regard to the establishment of tariffs, funding, personal policy etc.

The activity which we have outlined would include more tasks than that of today. It should be concerned with local energy sources such as sun, wind, small hydro power plants, combustion or pyrolysis of refuse, etc. It must be capable of being coordinated with adjoining local authorities. It should also include the planning of energy conservation in the local government area. The energy utilities ought perhaps to act in certain respects as middlemen between the credit market and the property owners.

These new tasks and wide powers pose certain demands. One of these is the demand for local political control and som - far from simple - coordination with other local government functions (planning of housing, traffic, etc). Their increased importance at a general political level might justify the setting up of special boards for energy management or possibly for energy and resource management. Should parliament institute an "act for local government energy boards" at the beginning of the 80's?

The strategic aims of the state

We have now set out the role of local governments in the energy policy of the future. But the major task of replacing oil without losing freedom of action cannot be left either to local governments, private organizations or companies. Compared to earlier transitions from one kind of energy to another, e.g. from coal to oil, the one with which we are now faced is more difficult. Most alternatives are more expensive than oil today. Many of them are in addition more awkward to handle. The conclusion is that the replacement of oil must be planned and a national strategy must be evolved for this.

The government and parliament must have the final responsibility for this strategy and ensure that the organization and framework of regulations act together to produce the desired result. It is equally clear that the state neither can nor should take the initiative in a number of detailed questions. This responsibility as in other spheres, must rest on those who are closest to the problems.

We shall deal below with four aspects of the state's responsibility: the replacement of heavy oil, the system of regulations for local governments and companies, consumer policy, and the public sector as a customer for technology.

A national strategy presupposes both measures for conservation and measures to exchange oil for other energy sources (see chapter 6). But the problems are fundamentally different for heavy oil and light fuel oil. The latter is used in a very large number of indvidual boilers, and it must therefore by and large be the task of local authorities to achieve the transition, facilitated and stimulated of course by central planning.

Heavy fuel oil (about half of the present-day oil consumption) requires different measures. It is used in a limited number (a few hundreds) of large installations. There are two ways to proceed - each of them with its own problems.

One is through the price mechanisms. An increased tax on heavy oil would stimulat users to save and convert to other fuels. Economic forces would be created to develop such fuels as peat and biomass and perhaps to import coal.

Another way would be for the state to control through direct regulations (concessions) how much energy and of what kinds could be used by local authorities and by energy-intensive industries. That is already done now in the case of heavy oil and coal, but for environmental reasons and not for reasons of energy management.

By gradually scaling down heavy oil through concessions and simultaneously expanding a supply capacity for other fuels such as coal, gas, peat, biomass, the state can control the overall fuel consumption. At the same time the financing of the investments which become necessary can be facilitated.

The strategy must thus be designed in such a way that a planned reduction primaril of oil (later also coal/gas)can be fully compensated by an increase in the supply of coal, gas, peat, biomass.

This is, in itself, a large task. To develop a capacity to supply, for instance, 10 Mtons of coal per year (corresponding to 70-80 TWh) requires contracts with coal exporters, possibly partial financing of mines, investments in harbour capacity, ships, and possibly rail capacity in Sweden, the establishment of coal stocks and handling systems for the user.

To extract 10 Mtons of peat per year also raises problems, of a partly different nature. Peat requires, among other things, extraction, collection, drying, transport and various new technical systems. Even if the peat-mosses are numerous and scattered, the systems development which will be required is probably such that a collective effort will be needed. Both Finland and Ireland undertake extensive peat extraction, in the development of which the state plays a decisive role.

We therefore regard concessionary legislation for the choice of fuel in processing industry, heating plants and power and heating plants as an important stage in a strategy to get rid of heavy oil.

The state has the ultimate responsibility for the overall system of rules and regulations (see section Energy enterprises and the complex of regulations in chapter 5). In our discussion of local government energy planning we noted that several of the laws which regulate the activity of local governments may need to be changed. That applies to laws concerning local government energy planning, local government commercial operations, and possibly the establishment of local government energy boards. The extended role of local governments would on the other hand involve the repeal of many of the regulations existing at present, for instance in the area of building legislation and building standards, and the transfer of powers of decision to the local government level. In short: a review of the regulations which determine the activity of local governments with regard to energy policy will be necesssary.

But it is unavoidable that the state should also take greater responsibility for certain other parts of the formulation of regulations. Many of the regulations which determine the design of the energy system are formulated, for instance, by subsidiary organizations or through direct agreements between enterprises. This applies not only to the establishment of tariffs (to which we shall return) but also, for instance, to quality requirements for electricity or petrol and temperature control for heating systems. Many of these regulations are necessary stages in the coordination of various parts of a productive system - electrical quality and electric motors, petrol quality and petrol engines, etc. Many of them have a very strong influence on which new techniques are accepted and which are rejected.

The state can only use standards and regulations to promote technical development. This may, for instance, concern the efficiency of central heating boilers or the petrol consumption of cars. This approach has been used by the National Environment Protection Board in order to stimulate new methods in the sphere of environmental protection. The Building Research Council has indicated that certain types of buildings could well be designed, e.g. with solar heating systems, even today. By demanding (in Swedish Building Standards, SBN) that swimming baths, for example, should be heated with sunlight, such a large market for solar heating systems would be created that the uncertainty for innovators would be reduced. Other such standards could apply to hot water installations during the summer, heating systems for summer cottages, perhaps also the electrical systems for summer cottages.

This leads us on to <u>consumer policy</u>. The energy consumption of households can presumably be influenced to a large extent by consumer policy measures directed towards producers of appliances, cars, buildings, etc. Unfortunately there is not yet much experience of the form such standards should have. It is difficult to formulate standards which permit a wealth of variety in production and do not prevent the emergence of new ideas.

The discussion about cars in the United States is of interest here. The standards there do not apply to individual cars but to the <u>average</u> sold by each producer. If a producer sells a number of large cars, he must also sell a sufficient number of small cars so as not to exceed the average. And if he does, he is forced to pay a fine.

There are interesting possibilities for combining standards or regulations with economic control measures. Car taxes which are adpated to petrol consumption are one such way. The sphere of household capital goods - cooker, refrigerators, washing machines, etc - can partly be controlled through the formulation of the housing loan ordinance in such a way that only products with certain energy specifications would entitle to a loan.

There are further problems with standards as instruments of control, however. A large proportion of consumer goods are sold in an international market and it is possible that foreign producers and states would not stand idly by while isolated Swedish measures were adopted to promote consumer products using less energy. The whole area should therefore be studied more carefully and international cooperation should be aimed for.

To stimulate and control <u>technical development</u> in the energy sphere is a strategic task. The instrument which has hitherto been used and discussed is, above all, the distribution of resources for R&D. The very fragmented structure constituted by 278 local Swedish authorities is not sufficiently strong to promote technical development. Even if there are strong reasons in favour of local governments engaging in R&D more extensively than at present the responsibility must rest with the state. But R&D resources alone are not enough. There are reasons to doubt the gains of R&D unless it can also be shown that there is a likelihood of selling the relevant commodity. We therefore wish to indicate the possibility, as a complement to such investments, of consciously stimulating a market and thus promotin the development of new technical projects.

Such efforts do, in fact, have a long tradition in Sweden. The collaboration between ASEA and the State Power Board with regard to the development of systems for long-distance transmission of electricity is one example. The role of ASEA was to develop systems, that of the State Power Board to establish the requirements. Other examples are the Telecommunications Administration and LM Ericsson, the Aeronautical Equipment Administration and SAAB, etc.

The division of roles has thus been that the state has purchased complex technical systems and private or other suppliers have developed the new system in close collaboration with the procuring authority.

The state should be able to assume such a role as buyer in the energy sphere as well. There is, for instance, nothing to prevent the state from instructing the National Board of Public Buildings to introduce into all government buildings, e.g. heat pumps, solar heated hot water systems, fuel cells. Thereby the state would guarantee a market for the new technical solutions and the possibly most essential element of uncertainty for potential developers would disappear. The state could also extend its role of buyer to the local government sector through the Government Purchasing Proclamation.

The financing of energy management

The future energy policy will make great demands on financial resources. In that respect the nuclear and solar options do not differ.

Table 34 shows an estimate of the investment needs. The investments in electricit gas and thermal plants amounted to about 7 thousand million kronor in 1975. The uncertainties are naturally great with both alternatives, particularly towards the end of the period.

Table 34. Rough estmate of investment needs in Solar and Nuclear Sweden (thousand million kronor per year at 1977 money value).

	1995	2005	2015
Solar Sweden	10	20	25
Nuclear Sweden	15	30	10

Added to this will be resources up to 1990 for energy conservation, extension of the distribution system, above all for district heating, and for the supply system These amounts are comparable to the housing investments, which at present amount to 13 thousand million kronor per year, and they are undoubtedly so large that the capital market will be hard pressed. As other community functions also demand large amounts of capital, the capital market will have a major influence on the kind of energy supply which we will in fact obtain.

The capital market is at present organized in such a way that certain investments enjoy lower capital costs than others (see the section "Tariffs and finance as conditions for competition" in chapter 5). Measures which have the same effect in terms of energy policy (supplied or saved kWh) per invested krona cost very different amounts to investors. Investments in housing, which is financed through the housing loans system, undoubtedly enjoy the lowest capital costs. Next comes power production in which the State Power Board has a small lead over the private enterprises. District heating and co-generation are in theory in the same position as power production, but in practice in a worse one, as the Bank of Sweden does not grant permission for bond issues to the same extent in their case. The combination of energy-saving houses and direct electric space heating is thus financially favoured in relation to almost every other combination.

Investments in industry are subject to different conditions. Certain small investments have a repayment term of 1 to 2 years, largeer ones of 3 to 5 years, and more extensive ones of perhaps 10 years. In all cases the period is considerably shorter than that applying to energy production. Households also have an evident need for capital, even if it is irregularly distributed. Loans to households are usually the first to be tightened up duing periods of credit restrictions.

How will the Bank of Sweden open the gate of economic power for the various energy alternatives? The financing conditions strongly affect the costs of energy-saving and different techniques. (Photo: Pressens Bild)

This leads us to ask now only how much capital can be made available for energy mangement but also on what terms it should be offered to the various parties involved. There is no reason why a household that invests in a less energy-consuming refrigerator should have to pay a higher capital cost for the energy saved than the State Power Board would have to pay to generate it. It is a question of designing a system for the distribution of capital with the label "only to be used for energy-saving purposes", so that it is really used to the best purpose from the viewpoint of the overall energy management.

The conditions for different investments are determined not only by the position of an energy enterprise in the capital market but also by the degree of self-financing in the enterprise. Here the conditions differ above all between local government enterprises and national ones (power producers). The principle is that depreciation is based on the replacement value, which favours energy enterprises with a large proportion of older but still functioning installations (particularly old hydroelectric power plants) over energy enterprises with newer installations. The higher inflation is, the stronger is the effect of this. The systematic result is a distortion to the disadvantage of the local government energy boards. If this is to be corrected, a change of regulations is required with regard to investments, depreciation and tariffs.

The question of the form of the conditions for financing is thus connected with the form of tariffs. We have previously mentioned that the present-day principle of cost-related tariffs systematically distorts the choice of new technologies in favour of those with the same cost structure as the existing ones. A tariff system which favours new technologies, e.g. the technology of the solar option (or conservation), does on the other hand involve considerable financial risks for the present-day energy enterprises. In the electric power sector the picture is complicated by the fact that certain enterprises presumably run a greater risk than others, due to the relationship between new and old investments. If such contracts between bulk power producers and local authorities as stimulate co-generation plants in the local authority areas were to be made a general requirement, this could cause considerable economic problems for certain enterprises and large profits for others. The principles on which contracts and tariffs are based depend on the organization of the power industry and the division of roles between, for instance, the State Power Board and the private power producers.

The conclusion is that concentrated control is required over the economy of the energy sector, embracing both state authorities, private enterprises and local government organizations. The principles for establishing tariffs, regulations for amortisations, etc, must be reviewed and regulated in a different way than they are today, partly also in greater detail. It is also legitimate to ask whether there is not a need for a new form of capital supply. Perhaps a new "intermediate institution" which would be responsible for the whole of the energy-politcal part of the capital market?

The organization of the electricity sector

The discussion which we have conducted so far has indicated that the development of the electricity sector is decisive for the openess and ability to innovate which will be necessary if freedom of action is to be preserved.

The electricity sector must therefore be reorganized in order to respond to the following demands among others:

- planning of electricity production and heat production to be integrated
- electricity distributors promote electricity-saving
- local government energy boards and industrial enterprises are encouraged to produce their own electricity, e.g. through wind power, combined production etc
- tariffs are established which encourage conservation and new energy sources.

A completely new organization of the electricity sector naturally requires detailed studies. We shall only indicate here some of the main features.

One change of fundamental importance would be to make electricity distribution as a whole a primary local government task and to integrate it with the functions of the local government energy boards. Thereby the coordinated planning discussed above will be made possible.

The fixing of tariffs between power producers and local government electricity boards (and also large industrial enterprises) should stimulate local production by the latter. We have previously noted that such tariffs ill suit the interests of the power producers. One way of solving these problems would be to create a special state-owned wholesale enterprise for electricity ("state electricity board") separate from the existing production enterprises. Such a wholesale enterprise for electricity would be responsible for the transmission line grid (down to perhaps the 100 kV level), would buy electricity from the producers and sell it to local authorities/industry. The wholesale electricity enterprise could then pay for purchased electricity by and large at production cost and sell it at a price suited to stimulate local production by local authorities and inudstry in order to develop another kWh of bulk power. The wholesale dealer can choose between tenders for further electricity production, and present producers would have to compete with local authorities having surpluses.

Any profits of the electricity wholesaler can be funded and appropriately made available to local authorities/industry for electricity conservation or electricity-producing investments. The wholesaler would then function as financier and price regulator.

Several, though not all, of the tasks of the "state electricity board" are at present performed by the State Power Board. What will remain when distribution and the transmission line grid have been detached from that organization and transferred to the "state electricity board" will be a more straightforward nationalised commercial enterprise for electricity production. That would be a fairly logical separation of the two roles of the present State Power Board. The "state electricity board" would take over from the State Power Board the national role of promoting an efficient electricity system for the whole country. It should thus be regarded primarily as an instrument for policy.

8 Choosing a Future - Uncertainties and Values

Introduction

That the future is uncertain everyone knows. That one ought to take into account uncertainties in political and economic activity is also fairly generally accepted if not in practice then at least in theory. Problems arise when opinions differ on the extent of the uncertainty, how easy it is to reduce the uncertainty, and when it is time to make decisions.

We have outlined two possible and deliberately simplified principles for future Swedish energy supply. There are many dimensions involving great uncertainties today. Some of them are purely technical/scientific. Will it be possible to meet the material specifications which are required in order to introduce breeder reactors with such a short doubling time that the uranium needs will not be noticeably affected? Can energy forest plantations be made ecologically acceptable? Such uncertainties can often be reduced through more research and development.

Other uncertainties are more purely economic - what will breeder reactors, energy forest plantations, solar cells etc cost? Such questions are already more problematic - if we are to be able to assess, e.g. the kWh price of the output from breeder reactors and solar cells it is necessary to know what the production costs will be - and then one is dealing with yet another dimension of uncertainty. In that case it is not only the technical problems of materials in the breeder reactor which are uncertain, but also the feasibility of developing a production system which will produce material with specified qualities at a certain cost. The costs of solar cells are still more illuminating, as the general opinion appears to be that they can be reduced if the volume of production is sufficiently large, which in its turn would presuppose that the market is sufficiently large, which presupposes that the costs are sufficiently low. This becomes a vicious circle and the uncertainty about the future cost can only be resolved by starting production, which implies that one already has <u>faith</u> in the relevant technology.

This is in fact one of the focal points in the energy debate. It is hard to imagine that the development of nuclear power during the 50's and 60's - which did not really give concrete results in the form of electricity until the 70's - would have taken place unless those most closely concerned had believed in the idea and had faith in the chances of overcoming the difficulties which would arise. The same conditions, of course, apply now. Whichever way Swedish society chooses to go there will be a large number of difficulties - technical, environmental, organizational, etc. There is no doubt that the fundamental approach to them is itself of great importance. Are the material problems of the breeder reactor looked upon as a challenge or as an insurmountable obstacle? Is the requirement of reduced costs for solar cells seen as a challenge or as an impossibility?

The view of what is possible applies not only to the technology itself, however, but also to the people and organizations which will have to introduce and maintain it. Nuclear Sweden, for instance, will require an extremely competent central organization to develop and manage a fuel cycle based on plutonium. Will Sweden be able to maintain the competence of that organization at a sufficiently high level for a sufficient length of time? Solar Sweden poses similar questions. It presupposes that a number of industrial branches and professional groups learn quite new skills and that a shift takes place in the division of roles between the state, local authorities and private interests. Can these changes really be brought about?

The uncertainties in these respects are quite as large as the purely technical ones. And here, too, the fundamental approach to difficulties will itself play a decisive role, an approach which cannot be separated from what is seen as desirable and worth striving for. For that which people see as desirable they are also prepared to mobilise all their strength, but for that which is not desired one can only too easily convince oneself that it is impossible to achieve.

It is also absolutely clear from the energy debate up to now that people are not prepared to see future energy policy as a bundle of purely technical and organizational questions. Value judgements, faith and hopes about the development of society must decide in the final analysis. It is our main task in this study of the future to provide starting points for such basically political judgements, not to make them ourselves. But we would still like, in conclusion, to formulate a few ideas - primarily as raw material for our readers - about the wider questions which are raised when the uncertainties are to be transformed into decisions and reality.

Technology and societal structure

At one level this study concerns technology. The technical systems which today provide us with energy must be exchanged over a number of decades and there are various other technical systems that may be considered as replacements.

At another level the study concerns the interdependence between technical and institutional changes. When we change technology, we shall also chnge our institutions, and vice versa. Presumably only a fully nationalised electricity sector will be able to develop Nuclear Sweden complete with fuel cycle and all. Presumably only a mixed local government/state organizational structure can develop Solar Sweden.

But the effects of technology are more far-reaching than that. Technical changes are also correlated with social ones - sometimes extremely profound. Maybe technical change can also be seen as an instrument for social change?

An interplay between technology and social change has long existed in the energy sphere. Perhaps the clearest analysis was made long before the oil crisis in the mid-50's. Cottrell (1) then pointed out that, e.g., the introduction of draught-animals as an aid in agriculture not only made the work of human beings easier but also led to the production of a surplus. That made possible other activities than agriculture - specialised commerce, warfare, religion etc - and thus new forms of social organization. Many have traced the origin of civilisation to the fact that rivers like the Nile, Euphrates, Tigris, Indus and Yangtse-Kiang provide a surplus in the form of nutritious soil but also require an irrigation system and canals, which in their turn require organization. Cottrell points out that other energy sources than running water have also been significant. The sailing ship, for instance, can be seen as a technique for the harnessing of wind energy which was superior to other principles of producing kinetic energy (e.g. with slaves). The result was that a larger surplus could be created, voyages could be made over longe distances and the opportunities for trade increased. That brought with it specialisation and new professional groups.

White (2) points out that the transition from the ox to the faster horse in medieval England made it possible to loosen the connection between the place of abode and cultivated land. The transition from ox to horse involved a more efficient conversion of energy stored in plants to animal energy. That led to fundamental social changes and had repercussions in many spheres.

An example given by White is the effect on the art of warfare. We reproduce that example here with two purposes in mind. One is to illustrate that many activities in society have repercussions in quite different spheres. The other is the great difficulty of understanding in advance what all those consequences may be. The need to attempt in advance to establish and evaluate second and third order effects has been keenly appreciated in recent years. The discussions about technology assessment are a sign of that (3, 4, 5).

The English army under Edward I used the longbow as a weapon. In the hands of well trained archers it had a superior effect compared to crossbows. The English army was generally unbeatable right up to the end of the Hundred Years' War (1337-1453). The supply of skilled archers was sufficient during almost a century. In the middle of the 14th century it began to dwindle as a result of changing leisure habit

In 1365 Edward III ordered all sheriffs to ensure that bowls, handball, football, cock fighting and other "useless" games ceased and that Englishmen of the lower classes devoted themselves to archery in their spare time. In 1388 tennis and dice-playing were added to the banned games. As late as the 16th century similar measures were taken. This did not, however, prevent the decline in English archery skills. At the end of the 16th century the longbow was officially replaced in the English army by the musket, although still technically superior to the musket. The latter was relatively effective even in the hands of untrained soldiers.

What factors led to the change in leisure habits and thus in the long term to the musket replacing the longbow in the English army? There are presumably several mechanisms, but a significant one was that the horse gradually replaced the ox in agriculture. Because the horse was faster, it halved the time it took to get from farm to field. The peasants could therefore settle in larger villages instead of, as before, in separate farmsteads or smaller groups of houses near to the fields. Archery is a one-man sport and was therefore suitable in villages which were too small to accommodate an inn. As the villages grew largers, various kinds of team games also began to spread despite the attempts to ban them by decree.

The industrialisation during the 18th and 19th centuries can illustrate the interplay further. Industrialism is often associated with technical inventions like steam engines, but that is only half the truth. The steam engine could never have driven any textile machines if there had not been factories in which workers were gathered in one place. And the factories arose from the wish of the owners of capital and merchants to control the production process. Previously production often took the form now called the putting-out system, but the uneven quality and inadequate control led the owners of capital to become factory-owners. That provided a basis for mechanisation, mechanical operation and steam engines. Technology not only creates an economic and social order, it is also itself created by the social order.

It is obviously not easy to distringuish between cause and effect in the interplay between technical and organizational change. Yet one may wonder whether it is not the energy techniques - first the stationary steam engine, then railways and the steamships, electricity, the electric motor and the combustion engine - that have had the most fundamental influence precisely on social organization and the division of labour. Advanced specialisation presupposes geographical separation and the transport of people, materials and energy. High energy density - the amount of kWh/kg or litre - has presumably been necessary for this development.

The division of labour has continued. The concentration of energy turnover to increasingly specialised large-scale units seems inevitable.

It is not only the large energy enterprises - oil companies, electricity corporations, etc - which depend on high energy density and concentrated production. Processing industry, cities and transport systems have all been constructed around energy carriers of a high density.

At the same time people's anxiety appears to grow. Can one trust the oil companies to make real efforts to prevent "blow-outs"[1] from off-shore fields[2], or the nuclear safety, reprocessing, waste? Can one trust that the authorities appointed to look after the public requirements for protection of health and environment will not gradually come to regard themselves as part of the "energy industrial complex" rather than as its supervisors?

The anxiety surely derives partly from the feeling that the energy supply system has grown so rapidly, has become so large and complex that an overview of it is no longer possible. Oil supply, for example, consists of a gigantic international technical system for the extraction, transport, refining and retail distribution of oil. Nuclear power and its fuel cycle are similarly a technically complex system which is run by a number of experts and a large number of enterprises, agencies and governments. Is there really anyone who has an overall grasp of these tremendously complicated technical systems? Who bears the responsibility? Whom can we make accountable - if we should wish to?

Here the energy debate slides over into another debate, namely the one about trust and legitimacy in modern society.

Faith in the social system

Why do certain people regard one energy policy development as desirable but not the other? This is in itself an essential question, well worth deeper study. A study carried out by H Otway and K Thomas for IAEA/IIASA[3] (6) can here illustrate an idea which we ourselves believe to be worth further investigation.

They charted by means of a sociological method the way in which different attitudes to nuclear power develop. In the group that is chiefly "against" nuclear power the factors which most strongly influence attitude are in the following order:

(1) "Blow-out": When one reaches oil- or gas-carrying strata during drilling, the pressure can increase so fast that an uncontrolled escape of oil or gas takes place.
(2) Off-shore field: Oil or gas fields under the seabed.
(3) International Atomic Energy Agency, Vienna, Austria - International Institute for Applied Systems Analysis, Laxenburg, Austria.

1. Psychological motives: being exposed to risk without the ability to do anything about it, individual's lack of any possibility of exercising control.

2. Social and political motives: becoming dependent on elite groups and experts, being forced to depend on the stability of social institutions.

3. Environmental and health motives

4. Economic and technical considerations

The last group is allotted very little significance by those who were "against" nuclear power. Among those who were by and large "for" nuclear power, on the other hand, the main emphasis was on economic and technical advantages and, thereafter, on the environment and health.

It should be noted that what heads this list are two clearly political factors. They concern control and confidence.

The investigators also show how difficult the debate is. It is not really about different answers to certain questions but rather about what kind of questions should be regarded as significant. The authors summarise:

"In general, technologists have been surprised at the strength of public reactions against nuclear energy. Social scientists, in turn, have been equally surprised at the reluctance of technologists to accept the validity of public concerns about social issues as being quite distinct from the technical realities. It has been said that the fundamental concept in the physical sciences in energy and, in this same sense, power is the fundamental concept in the social sciences. If there is a central issue in the nuclear controversy it is personal and political power and public participation in the control of that power".

Some studies on confidence are of importance in this connection. The United States public, for instance, has adopted an increasingly cynical attitude to large corporations and governments. Distrust has grown rapidly. And even if much has happened - racial disturbances, the Vietnam war, Watergate - which is specifically American, the picture is by and large the same in other countries, including Sweden. In one study (7) it is noted that:

The number of voters who have deep confidence in parties is decreasing. The outward signs of harmony and trust are misleading. Under the calm surface there is among the body of voters an undercurrent of increasing alienation.

If many studies indicate that the distrust of the governors by the governed is increasing, how do the former regard the latter? Yankelovich (8) compared the view of the public held by professional groups and of the former's attitudes to energy-saving after the oil crisis and found that:

"For a large segment of the American public, perhaps a majority, we see a readiness evoked by the crisis to change and adapt, and a flexibility with regard to the future that is almost universally underestimated by the professionals we interviewed in the study. These observers tended -- as often happens with elites -- to impute flexibility, imagination, generosity of impulse and breadth of understanding to themselves, and rigidity, lack of imagination, and narrow self-interest to the general public".

If Yankelovitch is right about the views held by the United States elite about the United States public, this provides cause for thought. Could we become victims of what somebody has called "greed by proxy"?

It is presumably a good deal more important for the survival of society that faith in the political system and the institutions of society ceases to decline and instead increases than that GNP grows by a slightly higher or lower percentage. The question is only whether there is a connection between them. We shall briefly touch on that in the next section.

Energy and everyday life

Energy in everyday life is not only warm houses, time saved through cars, washing machines and vacuum cleaners, municipal swimming baths and voyages to Ceylon. Nor is it only current for the lathe at one's work, the oil that heats the cement kiln, or the electric power in the aluminium factory. It can also mean polluted beaches in the archipelago, increased cancer risks from car exhaust, demands for land to carry 800 kV transmission lines, gashes in the Billingen area and devastated farming country along the Kalix river. Added to that is the image of giant oil corporations of which several have a turnover which is larger than Sweden's national budget. There may also be a feeling of powerlessness in the face of the complexity itself.

All this and more has to be taken into account when one tries to assess whether people accept of appreciate moderation in material standards in order to attain other values. In this section we shall look at various surveys of opinions and attitudes around these and related issues.

According to the post-election survey of the Central Bureau of Statistics (7) there is a significant majority in Sweden in favour of conserving energy even if that means a fall in the standard of living. This majority in the population as a whole is 70%, and clearly preponderant in all political parties. There is also a definite majority in favour of investing in new energy sources and opposed to nuclear power, hydro power and oil imports.

Other studies show a similar result. According to one study by SIFO a clear majority of the Swedish national is prepared to invest more in environmental protection and health services and regards that as more important for the future than a continued growth in material standards (9).

Such feelings appear to be part of a current change of opinion in several industrial countries. That is demonstrated by comparative studies at different points in time within one country as well as between different countries. There seems to be a growing determination to place environment, intellectual freedom and mobility, less impersonal societies, co-determination at the workplace and a consumer-oriented economy above, for example, high economic growth and a consumption-oriented economy. This determination seems to be relatively independent of the level of economic development (measurements have been made both during the boom at the beginning of the 1970's and the slump in 1973). There are, on the other hand, differences between age groups. Those who have experienced depression and instability during the 30's and second world war appreciate a stable growth in consumption more than younger people who tend to take the post-war increase in affluence for granted (10).

Some have even spoken of a new ethic, a "conservation ethic":

"This is already happening. For example, in the United States we are now witnessing a transformation of the success ethic. Instead of focusing exclusively on status and material possessions, the definition of success is being broadened to include the concept of the 'full, rich life'. The full, rich life includes possessions, but it also emphasizes intangibles relating to health, friendship, exercise, communion with nature, physical well-being, ethnicity, creativity, self-knowledge, social skills, sexual expression, self-fulfillment and adoption of pluralistic life styles. This enlargement of the idea of success can far more readily be assimilated into and integrated with a new conservation ethic than could the old conception of success tied narrowly to possessions and consumption". (8)

There thus seems to be a willingness to forego further economic growth in order, instead, to gain in other qualities. It also seems as if this willingness is connected with the security which a steady economic growth has in face provided. The question is therefore whether the willingness to forego growth presupposes a security which in itus turn presupposses growth. In that case the picture becomes more complex.

The debate about the disadvantages of growth is new and can be dated to the middle and end of the 60's. But the view of economic growth as something desirable and attainable is also fairly new - it largely emerged after the war.

There is now a tendency to ascribe to economic growth many of the bad aspects of society: destruction of the environment, exhaustion of resources, loss of employment, isolation in housing areas, etc. There is also another current which maintains that it is only through increased growth that we can create resources for environmental protection, social care, housing construction, automation of oppressive work, etc.

It is not very easy to unravel the essentials in this discussion, but we still do not belive that the problems referred to can be solved by means of economic growth. It is therefore almost equally unjustified to argue that we must have contined growth in order to solve such problems as to assume that, if growth can be planned down to zero, the problems would disappear. It is instead more complex political problems that are being posed.

In the study to which we referred earlier, Yankelovich (8) studied the United States, West Germany and Sweden after the oil crisis in order to find out how the general public on the one hand and the professional, dominant elite on the other looked at energy, lack of resources and living standards. He found

that Swedes and West Germans are able to a greater extent than Americans to accept that material standards did not rise

that Swedes lay greater emphasis on a good health service, secure pension arrangements and full employment than do West Germans and Americans

that people in the United States and West Germany with higher incomes find it easier to accept a limitation of the rise in standards than people with lower incomes. In Sweden the opposite actually applied

that the level of sacrifices which people can accept is clearly dependent on how equitably they judge these sacrifices to be distributed

that the willingness to accept sacrifices also depends on the sense of participation, co-determination and alienation.

The consequences of this may be very fundamental. If the government, parliament and the profession elite regard it as necessary to make demands on citizens to reduce their rise in standards or to reduce their standards, then such a policy must have certain definite characteristics. The citizens pose counter-demands. They are more prepared to accept such a policy if it appears to affect everyone relatively equally and, even more so, if it increases the sense of justice.

Such a policy is not very easy to realise. Certain privations are fairly easy to share out evenly among the population - e.g. reduced purchasing power in the form of of a higher value-added tax or higher energy taxes. Others, which would also follow from a change to higher energy prices, are not. If the energy policy contributes to inflation, some will in practice be favoured more than others. The closure of factories hit those who work there very hard, but the citizens in general are only very indirectly affected. The requisition of land for wind power generators, power transmission grids or nuclear power reactors will inflict limited sacrifices on certain groups, while others chiefly gain advantages. If the geographical and administrative gap between production and use is small, the advantages and disadvantages may be easier to relate to each other and therefore to accept.

Freedom for one person can easily become a restraint on another. This applies to concrete things such as how the land is to be divided between recreation, forestry and energy plantations, but it also applies to basic social values such as influence and participation. Of course the freedom to choose a form of heating may be important. But if many choose direct electric space heating a situation will gradually be created where there is no alternative, e.g. to nuclear power; than the greater possibilities of choice in one area have created limited possibilities of choice in another.

The readiness of people to accept an overall view does, we believe, depend on whether they have trust in the people and the laws which regulate the course of events. Democracy will be the decisive factor.

The society of the future and the alternatives

In this chapter we have indicated even more fundamental problems and uncertainties of different future alternatives than those directly connected with technology and economics. What demands does the energy system make on society? How are social evolution and public spirit affected by the choice of energy system? Here the gaps in our knowledge of mechanisms and connections are evident and the uncertainty is of such a nature that one simply cannot get a definite answer in advance. Fears and hopes become more an expression of a general view of society.

The difficulty involved in not having access to the answers must not, however, prevent a discussion of the problem. Many regard these aspects as the most decisive reasons for going one way or the other. Lovins (11), Hayes (12) and others have thus held that a nuclear power option necessarily leads to a very centralised and elitist society while the solar option, on the other hand, would lead to a society which in several respects would be more decentralised. Yet others have warned against the detailed regulation and the great amount of state influence - an ecological "big brother" - which would be necessary to carry out the adaptation required by Solar Sweden. The neeed for security measures to prevent plutonium going astray has led many to conjure up a picture of Nuclear Sweden as a totalitatian police state. And even without going quite so far, some of the advocates of nuclear power have spoken of a "Faustian choice"; we can have access to large amounts of energy from breeder reactors but only if we commit ourselves to social institutions which can cope with the technology and which will have a life-span far beyond what mankind has hitherto been used to (13).

We shall finally speculate on Nuclear and Solar Sweden from the perspective to which we have referred earlier in the chapter. What social relations and values can be strengthened and facilitated if we move in the direction of Nuclear Sweden and which ones if we move in the direction of Solar Sweden? We would stress that we do not believe that the choice of energy technique will <u>create</u> social relations - but on the other hand the choice can <u>reinforce</u> tendencies that already exist.

An energy system based on ncuelar power has certain definite characteristics. It uses an energy carrier (electricity) which is extraordinarily flexible and convenient - the consumer gets all the energy he needs from two holes in the wall without having to directly concern himself about what is on the other side. This is indeed one of the greatest and most complex technical systems hitherto created by mankind. A system that has not made any other demands on the individual than that he should accept certain inconveniences in connection with the siting of installations, restricted access to certain areas, etc. The combination of comfort and inaccessibility is underlined by the fact that the qualifications demanded to manage the energy system are highly specialised.

The institutions required by nuclear power to deal with siting, the capital market, management of the electricity system, dependence on international technology etc in Nuclear Sweden will lead to a profound separation between producer and consumer. The latter will have to pay for his comfort by abandoning the demands for participation and a sense of co-responsibility.

The tariff arrangements, among other things, will reinforce this tendency. Individual thriftiness may appear virtually as an uncontrolled disturbance and threat. It is easy to imagine that that will reduce rather than increase the sense of participation.

Such conditions could lead to a reinforcement of individualistic values such as: family-centredness, individual consumption, possibly a certain alienation from society.

The technology of Solar Sweden makes partly similar but partly different demands. Heating and parts of the electricity supply system will be based on installations dimensioned for blocks and urban sectors. The qualifications demanded are not so highly specialised as in Nuclear Sweden but relate rather to a large number of functions such as water, heating and sanitation technique, forestry technique, combustions technique, chemical processes, etc. The technical design will require organization at all levels. The state is not enough. There will also be a need for an organization to operate the systems and also a contribution from the user. Solar Sweden will be facilitated by cooperation at the urban block level.

The technology of Solar Sweden would appear to involve more cooperation and joint decisions on priorities than that of Nuclear Sweden. One can therefore guess that Solar Sweden would strengthen processes like integration, the reduction of gaps between producer and consumer, an increased emphasis on the role of the user. Against such values as greater comfort and individual freedom of choice, which will perhaps apply in Nuclear Sweden, there might in Solar Sweden be values like participation and co-responsibility, but also a certain compulsion to conform to one's environment.

Torsten Hagerstrand has suggested the concept "reach" to capture the need for coherence in the world of man (14). Grasp has not only a purely physical meaning but also economic, epistemological and emotional dimensions. It is a matter, among other things, of the need to understand something of the world and the technology which surrounds us. How many know how a nuclear power station or a wind power generator functions? Has such knowledge a value?

Bibliography and Notes

Energy - some methods of approach

1. M King Hubbert. The Energy Resources of the Earth. Scientific American 224 (1971), pp 60-87.
   F J Dyson. Energy in the Universe. Scientific American 224 (1971, pp 50-59.
   G Gustafson, S Lyttkens, S G Nilsson. Energy Forms in the Universe. The Energy Flow to Earth. In Energy - Not Solely a Question of Technology (in Swedish). Centre for Interdisciplinary Studies. Gothenburg 1974.
2. Goran Wall. Exergy - a Usable Concept in Resource Accounting (in Swedish). Department of Theoretical Physics, Chalmers Institute of Technology. Report No. 42-77. Gothenburg 1977.
   Svenska Dagbladet 75-12-06. Exergy - Useful Energy (in Swedish), p 3.
3. Efficient Use of Energy: A Physics Perspective. American Physical Society. January 1975.
4. Robert H Socolow: The Coming Age of Conservation. Annual Review of Energy. Vol 2, 1977.
5. Incomes of the Swedish People (in Swedish). SOU 1970:34 (and other publications up to 1977)
6. Sweden's Use of Energy in the 1980's and 1990's (in Swedish). SIND 1977:9. The National Swedish Industrial Board. Stockholm 1977.
7. Thomas B Johansson, Peter Steen. Secretariat for Future Studies. Stockholm 1978.
8. Lars Bergman, Harry Flam. Energy and Economic Growth (in Swedish). Secretariat for Future Studies. Stockholm 1976.
9. Bo Diczfalusy. Energy and Income Distribution (in Swedish). Secretariat for Future Studies. Stockholm 1976.
10. Situation report from Government Committee on Energy Forecasts (EPU). Ds I 1973:2
11. Mans Lonnroth. Energy, Employment and Welfare in Energy Policy for the Future (in Swedish). The Federation of Swedish Industries. Stockholm 1977.
12. Thomas B Johansson, Mans Lonnroth. Energy Analysis - An Introduction (in Swedish). Secretariat for Future Studies. Stockholm 1975.
13. Lars Bergman. Energy and Employment (in Swedish). Stencil. Secretariat for Future Studies. Stockholm 1976.
14. A Carling, J Dargay. Energy Consumption of Households - A Study of the Effect of Energy Taxes on the Distribution of Real Income. Annex 16 to "Policy Instruments for Future Conservation of Energy" (in Swedish). Main report from an Expert Group to the Energy Commission.

15. Cf Erik Allard. To Have, To Love, To Be. On Welfare in the Nordic Countries (in Swedish). Argus, Lund 1975.
The Secretariat for Future Studies arranged during its first year of activities a symposium to illustrate different aspects of quality of life. This is reported in Quality of Life (in Swedish) (Ds Ju 1974:9). Liber forlag. Stockholm 1974.
16. A Study by Kerstin Elmhorn in cooperation with the Secretariat for Future Studies will illustrate the differences in use of energy for different life styles.
17. Energy Commission Policy Instruments for Future Conservation of Energy. Ds I 1977:15.
18. Cf Lars Lundqvist. Market Economy and Zero Growth in the Use of Energy (in Swedish). Department of Mathematics, Royal Institute of Technology. PM 1977.
19. CONAES - Committee on Nuclear Energy and Alternative Energy Systems. Project at National Academy of Sciences, USA. The project is to be completed during 1978.
20. P G Kihlstedt. National Raw Materials Supply during Energy Shortage (in Swedish). Academy of Engineering Sciences (IVA). Report No 112. Stockholm 1977
21. Alf Carling, Joyce Dargay, Christina Oettinger, Asa Sohlman. The Effects of Energy Policy in Industry.
Annex 14 to Policy Instruments for Future Conservation of Energy. Ds I 1977:17 Ministry of Industry, The Energy Commission.
22. Alf Carling, Ds I 1977: 11.
23. L Emmelin, Bo Wiman. On Energy and Ecology (in Swedish). Secretariat for Future Studies. Stockholm 1977.
24. Per Anders Bergendahl. Structural Changes and Energy Consumption (in Swedish). Research Group for Energy System Studies, Department of National Economy. Stockholm University. Publication No 1978:1. See also P A Bergendahl, C Bergstrom, A Carling, A Sohlman, G Ostblom. Energy, Structural Transformation and Employment. Energy Research and Development Commission (DFE). DFE Report No 9. (In Swedish)
25. Power Development 1975-1990. Central Operating Management. A Joint Organization of Major Swedish Power Producers (CDL). Stockholm, September 1972. (In Swedish)
26. Government Committee on Energy Forecasts. SOU 1974:64. Energy 1985-2000 (In Swedish)
27. Energy, Health, Environment (in Swedish). SOU 1977:67.
28. Energy Commission, Group on Safety and Environment.
29. Lars Emmelin, Bo Wiman. Effects of Energy Production on the Environment (In Swedish). Secretariat for Future Studies. In press.
30. Nils Malmer. On the Effects on Water, Soil and Vegetation of Increased Sulphur Supply from the Atmosphere (In Swedish). The National Swedish Environment Protection Board. PM 402, October 1973.
31. I T Rosenkvist. A Contribution to Analysis of Buffering Properties of Geological Materials against Strong Acids in Precipitation (in Norwegian). The Norwegian Scientific Research Counil 1976.
32. P Grandjean, T Nielsen. Organic Lead Compounds - Pollution and Toxicology (In Norwegian). Report from the National Swedish Environment Protection Board (SNV). PM 879, 1977.
33. See, for example, Ingvar Bergkvist, Bengt Hansson, Holger Rootzen, Tord Torisson. With what Assurance Can One Know Anything about the Accident Risks in Complicated Technical Systems? (In Swedish). In Risk Evaluation, Ds I 1978:15. The Swedish Ministry of Industry, the Energy Commission and the project Risk Generation and Evaluation in a Societal Perspective. Committee for Future Oriented Research, Fack, S-103 10 Stockholm.
34. Rolf Edberg. To Live on the Earth's Conditions (In Swedish). Documenta No 11, The Royal Swedish Academy of Sciences. Stockholm 1974.

Energy and societal development in interaction

1. Power Development 1975-1990. Central Operating Management. A Joint Organization of Major Swedish Power Producers (CDL). Stockholm, September 1972. (in Swedish)
2. Government Committee on Energy Forecasts. SOU 1974:64. Energy 1985-2000 (in Swedish)
3. The period 1800-1935. O Warneryd, A Jarnegren, F Ventura. Societal Development and Energy Supply (in Swedish). Stencil. Secretariat for Future Studies. Stockholm, December 1976. The period 1936-1950. Fuel and Power. Survey of Sweden's Energy Supply (in Swedish). SOU 1951:31. The period 1950-1975. Energy conservation etc (in Swedish). Bill 1975:30.
4. Statistical Year Book and Fuel and Power. Survey of Sweden's Energy Supply (in Swedish).SOU 1951: 32.
5. Facts about Oil (in Swedish). SOU 1953:12.
6. J M Blair. The Control of Oil. Pantheon Books. New York 1976.
7. Anthony Sampson. The Seven Sisters. The Great Oil Companies and The World They Made. Hodder & Stoughton. London 1975.
8. Sheldon Novick: The Auto. Environment, Vol 18, No 3, April 1976.
9. See, for example, Lars Lundgren. Energy Policy in Sweden 1890-1975 (in Swedish). Secretariat for Future Studies. Stockholm 1978.
10. Gustaf Olsson. Local Government Energy Supply - Technology and Politics in Cooperation (in Swedish). Vasteras 1974.
11. O Warneryd, A Jarnegren, F Ventura. Societal Development and Energy Supply (in Swedish). Stencil. Secretariat for Future Studies. Stockholm, December 1976.
12. Energy Conservation. Main report from an Expert Group to the Energy Commission Ds I 1977:10.
13. Lars Bergman, Harry Flam. Energy and Economic Growth (in Swedish). Secretariat for Future Studies. Stockholm 1976.
14. Calculated from Ds I 1977:12 and SOU 1974:64.
15. Bo Diczfalusy. Energy and Income Distribution (in Swedish). Secretariat for Future Studies. Stockholm 1976.
16. T Cronberg, I-L Sangregorio. Inside One's Own Door. (in Swedish). Secretariat for Future Studies and National Board for Technical Development. Stockholm 1978.
17. Anell et al. Shall we asphalt Sweden? (in Swedish). P A Norstedt. Stockholm 1971.
18. Lars Bergman. Energy and Employment (in Swedish). Stencil. Secretariat for Future Studies. Stockholm 1976.
19. Bo Carlsson. The Relative Trend for Energy and Its Significance for Energy Consumption, Industrial Structure and Choice of Technology - An International Comparison. Annex 12 to Policy Instruments for Future Energy Conservation. Ds I 1977:15. (in Swedish)
20. S Selander, The Living Landscape in Sweden (in Swedish). Bonniers 1957.
21. Lars Emmelin, Bo Wiman. On Energy and Ecology (in Swedish). Secretariat for Future Studies. Stockholm 1977
22. Fuel and Power. Survey of Sweden's Energy Supply (in Swedish). SOU 1951: 32.
23. Fuel Supply in the Atomic Age. Part I (in Swedish). SOU 1956: 46.
24. The Trade in Oil (in Swedish). SOU 1947:14.
25. Fuel Supply in the Atomic Age. Par II (in Swedish). SOU 1956:8.
26. Federal Energy Administration: The relationship of oil companies and foreign governments. Washington, DC, June 1975.

Sweden's energy supply from an international perspective

1. J Darmstadter, J Dunkerley, J Altman: How Industrial Societies Use Energy. Resources for the Future. Johns Hopkins University Press. Baltimore, USA 1977.

2. Handbook of international trade and development statistics. UNCTAD, Geneve, New York 1976.
3. Gosta Tompuri. Energy and the Development of Developing Countries (in Swedish). Secretariat for Future Studies. Stockholm 1977.
4. Parikh, Parikh: Mobilization and impacts of biogas techniques. IIASA RM 77-26. Vienna, November 1977.
5. Statistical Year Book 1975. The Swedish Central Bureau of Statistics.
6. World Energy Demand to 2020. Energy Research Group, Cavendish Laboratory, Cambridge.
7. Tor Ragnar Gerholm. WEC's Global Energy Studies (in Swedish). Address at "ENERGI 77" on November 1, 1977 arranged by Central Operating Management. A Joint Organization of Major Swedish Power Producers (CDL) et al.
8. W Haefele: Energy Systems: Global Options and Strategies. Institute for Applied Systems Analysis (IIASA). Vienna, May 1976.
9. E Eckholm: Losing Ground. World Watch Institute. W W Norton, New York 1976.
10. Denis Hayes: Rays of Hope. W W Norton, New York 1977.
11. Denis Hayes: Energy for Development: Third World Options. Worldwatch Paper 15. Worldwatch Institute, Washington, D.C., December 1977.
12. Arjun Makhijani with Alan Poole: Energy and Agriculture in the Third World. Ballinger Publishing Company, Cambridge, Mass. 1975.
13. National Academy of Sciences: Energy for Rural Development: Renewable Resources and Alternative Technologies for Developing Countries. National Academy of Sciences, Washington, D.C. 1976.
14. Energy Conservation etc. Swedish Government Bill 1975:30
15. OECD: World Energy Outlook. Paris, January 1977.
16. Peter Steen: On Oil Supplies (in Swedish). Secretariat for Future Studies. Stockholm 1977.
17. The Oil and Gas Journal: Worldwide report: Reserves, operations hit new highs. December 26, 1977.
18. D A Rustow: U.S. - Saudi Relations and the Oil Crisis of the 1980's. Foreign Affairs, April 1977, pp 494-516.
19. The Oil and Gas Journal: New Energy Sources Needed by 1990's (forecast from Exxon), March 28, 1977/pp 36-37.
20. L W Fish: Study for the Conservation Commission on World National Gas Supply. World Energy Conference 1977.
21. E Moberg: The World's Energy Supply during the 1980's and 1990's - Some Numerical Examples. Stencil. Secretariat for Future Studies. Stockholm 1977-09-30. (in Swedish)
22. For a summary account see B Bolin: Energy and Climate. Secretariat for Future Studies, Stockholm 1975, and Bjorkstrom, Bolin, Rodhe, Report to the Energy Commission, Group A.
23. Johan Schlasberg: To understand Coal (in Swedish). Secretariat for Future Studies. Stockholm 1978.
24. W Peters, H-D Schilling: An appraisal of world coal resources and the availability. A Discussion Paper prepared for the Conservation Commission of the World Energy Conference, August 1977.
25. W. Greider: U.S. Oil Industry Stakes Out Role for the Future. Washington Post. 1977-05-22.
26. Richard G Hewlett, Francis Duncan. Nuclear Navy 1946-1962. The University of Chicago Press. Chicago and London 1974.
27. Federal Energy Administration: National Energy Outlook - 1976. Washington D.C.
28. Alvin M Weinberg: To Breed or not to Breed? Across the Board. September 1977.
29. Studsvik Energy Technique, Inc, in Swedish Government Budget Proposals 1970, Annex 15.
30. The contribution of nuclear power to world energy supply 1975-2020. Report on Nuclear Power to the World Energy Conference 1977.
31. International Review: France 7705. Breeder reactors (in Swedish). Academy of Engineering Sciences (IVA). Stockholm, November 1977.

32. Nuclear power growth 1975-2000. Forecasts and projections. International consultative group on nuclear energy. Stencil May 1977.
33. Electricity Supplies in Sweden 1975-1990 (in Swedish). Central Operating Management. A Joint Organization of Major Swedish Power Producers (CDL). Stockholm 1972.
34. Sweden's Use of Energy in the 1980's and 1990's (in Swedish). National Swedish Industrial Board (SIND). SIND 1977:9. Stockholm 1977.
35. I C Bupp: Energy Policy Planning in the United States: Ideological BTU's in Leon N Lindberg (editor): The Energy Syndrome. Lexington Books. Lexington, Cambridge, Mass, and Toronto 1977.
36. Power Reactor Supply. Stencil May 1977. International Consultative Group on Nuclear Energy.
37. International Uranium Cartel. Volume 1. Hearings before the Subcommittee on Oversight and Investigations of the Committee on Interstate and Foreign Commerce. House of Representatives, May 2, June 10, 16 and 17; and August 15, 1977. Serial No 95-39. US Government Printing Office. Washington, D.C. 1977.
38 United States Atomic Energy Commission. Reactor Safety Study. WASH-1400 (NUREG-75/014).
39. Environmental Effects and Risks in Use of Energy (in Swedish). The Energy Commission. Stockholm 1977.
40. I Bergkvist, B Hansson, H Rotzen, T Torisson: With What Assurance Can One Know Anything about Accident Risks in Complicated Technical Systems? (in Swedish). Report to the Energy Commission, Group A. Stockholm 1977.
41. Report of the Nuclear Energy Policy Group: Nuclear Power Issues and Choices. Ballinger, Cambridge, Mass. 1977.
42. Bertil Persson, E Holm: Plutonium and Other Transuranic Elements: Radioecology Radiotoxicology (in Swedish). Report to the Energy Commission, Lund, June 15, 1977.
43. See, for example, article on laser development in Wall Street Journal, 14 December 1977.
44. D E Ferguson: Simple, quick reprocessing plant. Memorandum, August 30, 1977. Oak Ridge National Laboratory (ORNL).
45. Solar Energy, a UK assessment. Prepared by UK section of the International Solar Energy Society, The Royal Institution, 21 Albermarle Street, London W1X 4BS, May 1976, p 77.
46. Wind Energy in Sweden (in Swedish). National Swedish Board for Energy Source Development (NE). NE 1977:2. Liber. Stockholm 1977.
47. W D Metz_ Fusion Research I-III. Science, Vol 192, p 1320, Vol 193, pp 38, 307.
48. Energy, Health, Environment (in Swedish). SOU 1977_67.
49. Swedish Petroleum Institute. Annual Report September 1, 1976 - August 31, 1977 (in Swedish).
50. Personal communication from Ingemar Lundholm, Swedish Nuclear Fuel Supply Company.
51. OECD Nuclear Energy Agency and IAEA: Uranium resources, production and demand. December 1977.
52. J M Blair: The Control of Oil. Pantheon Books, New York 1976.
53. Anthony Sampson, The Seven Sisters. The Great Oil Companies and The World They Made. Hodder & Stoughton. London 1975.
54 Jerome Martin Weingart: The Helios Strategy. IIASA, Laxenburg. WP-78-8. May 1977, Revised February 1978.
55. Melvin A Conant et al: International Energy Supply: An Industrial World Perspective. Rockefeller Foundation. March 1978.

Two energy futures

1. P A Bergendahl: Energy and Employment (in Swedish). Stencil 1975.
2. Sweden's Use of Energy in the 1980's and 1990's (in Swedish). National Swedish Industrial Board (SIND). SIND 1977:9. Stockholm 1977.

3. Sweden's Energy Supply - Energy Policy and Organization (in Swedish). SOU 1970:13
4. Power Development 1975-1990 (in Swedish). Central Operating Management. A Joint Organization of Major Swedish Power Producers (CDL). Stockholm, September 1972
5. Industry in National Physical Planning - The Power Industry (in Swedish). National Swedish Industrial Board (SIND). SIND PM 1977:3. Stockholm 1977.
6. Sweden's Long-Term Energy Supply (in Swedish). The Swedish Academy of Engineering Sciences (IVA). IVA Communication 209. Stockholm, May 1977.
7. 65% or 5700 hours/year corresponds roughly to the operational statistics available today (8). A usage time of 70% is hoped for (9).
8. Jorgen Thunell: Nuclear Power in a Cul-de-Sac? (in Swedish). CDL/Ingenjorsforlaget. Stockholm 1977.
9. The Energy Commission Supply Group. Report Summer 1977.
10. The following fuel parameters have been used. Usage time 65% (5700 hours/year). Natural uranium requirement 27 tons/TWh electricity if no reprocessing; 18.9 tons TWh electricity with reprocessing and recycling of uranium and plutonium (The Energy Commission Supply Group, Report Summer 1977, Section 4.6). Using plutonium produced in breeder reactors, only the uranium obtained in reprocessing can be recycled to the light water reactors. For this case we have used a figure of 22 tons natural uranium requirement per TWh electricity.
11. The Energy Commission Safety Group. Preliminary Report, Part 2, October 2,1977.
12. See the Energy Commission Safety Group.
13. The Energy Commission Supply Group. Report Summer 1977, Section 4.6
14. Final Stage of the Nuclear Fuel Cycle (in Swedish). The Nuclear Fuel Safety Project. Stockholm 1977.
15. Future National Physical Planning. Study of the Power Industry, 1975 (in Swedish). Central Operating Management. A Joint Organization of Major Swedish Power Producers (CDL). Stockholm 1977.
16. Siting of Large Coastal Heat and Power Stations (in Swedish). Central Operating Management. A Joint Organization of Major Swedish Power Producers (CDL). Stockholm 1972.
17. The Energy Commission Safety Group. Preliminary Report, October 1977.
18. Industry in National Physical Planning - The Power Industry (in Swedish). National Swedish Industrial Board (SIN). SIND PM 1977:3. Stockholm 1977.
19. Estimated by assuming that about 70 km of grid is needed per GW electric power. This figure obtained by assuming that the 2100 km 800 kV suffice for a capacity three times the 1975 nuclear power programme (i.e. 30 GW).
20. Is Night Accumulation Economically Justified? (in Swedish). Swedish State Power Board. 1977-04-05.
21. Report of the Nuclear Energy Policy Study Group: Nuclear Power Issues and Choices. Ballinger, Cambridge, Mass. 1977.
22. Trends in Light Water Reactor Capital Costs in the U.S.: Causes and Consequences. Center for Policy Alternatives, MIT, CPA 74-9, December 18, 1974.
23. Rolf G Karlsson: Land for Energy Forest Cultivation. National Physical Planning (in Swedish). Basic data No 14-77. Ministry of Housing 1978.
24. Planning Report "Solar Energy in Sweden" (in Swedish). National Swedish Board for Energy Source Development NE). NE 1977:21. Stockholm 1977.
25. Wind Energy in Sweden (in Swedish). National Swedish Board for Energy Source Development (NE). NE 1977:2. Liber. Stockholm 1977.
26. Per Ragnarson: Renewable Energy Sources (in Swedish). Secretariat for Future Studies. Stockholm 1977.
27. Office of Technology Assessment: Application of Solar Technology to Today's Energy Needs. Washington, D.C. June 1977.
28. Wave Energy in Sweden (in Swedish). National Swedish Board for Energy Source Development (NE). NE 1977:4. Stockholm 1977.
29. Salt Energy in Sweden (in Swedish). National Swedish Board for Energy Source Development (NE). NE 1977:22. Stockholm 1977.
30. National Benefits Associated with Commercial Application of Fuel Cell Powerplants. United Technologies Corp, February 1976, Prepared for ERDA, ERDA 76-54.

31. Wind power (in Swedish). Swedish State Power Board. Stockholm 1974.
32. Energy Research and Development-78 - Energy Production (in Swedish). Prepared by the National Swedish Board for Energy Source Development. SOU 1977:61.
33. Taylor Associates. A preliminary assessment of prospects for worldwide use of solar energy. Report to the Rockefeller Foundation. Stencil 1978.
34. Lars Emmelin, Bo Wiman: On Energy and Ecology (in Swedish). Secretariat for Future Studies. Stockholm 1977.
35. For fuller account see Thomas B Johansson, Peter Steen: Solar Sweden. Secretariat for Future Studies. Stockholm 1978.
36. The 1975 Long-Term Planning Committee. SOU 1975:89.
37. Swedish Central Bureau of Statistics: Manpower Resources 1965 - 2000 (in Swedish). Information on forecasting questions 1976:1. Stockholm 1976.
38. Amory Lovins: Soft Energy Paths: Toward a Durable Peace. Ballinger 1977.
39. Estimate from Bo Diczfalusy: Energy and Income Distribution (in Swedish). Secretariat for Future Studies, Stockholm 1976, pp 8-9. The knowledge of the distribution of available income is scarce.
40. M Lonnroth, T B Johansson, P Steen: Energy in transition. Secretariat for Future Studies. Stockholm 1977.
41. For lack of knowledge as to how much additional power is needed to level off short-term fluctuations of demand, the following model was used: per installed MW nuclear power 0.4 MW is required for load management (pump power, fuel cells). This was obtained from a CDL study (CDL = Central Operating Management. A Joint Organization of Major Swedish Power Producers), (National Swedish Industrial Board, SIND 1977:3) by comparing the 250 and 200 TWh alternatives in the year 2005. A nuclear power increment of 7.5 GW in that study has led to a need for 2 GW gas turbine and 1.5 W pump power for levelling off the short-term variations (additional hydro power has also been installed, which is not taken into account here). Suitable locations for pumping power stations allow an output of about 5 GW (CDL study: Peak Power Stations in Swedish).
42. United States Atomic Energy Commission. Reactor Safety Study, WASH 1400 (NO-REG-75/014).

<u>Where are we going?</u>

1. Ingemar Dorfer, System 37 Viggen, Arms, technology and the domestication of glory. Universitetsforlaget, Oslo 1973.
2. See, for example, Donald A Schon: Technology and Change. The New Heraclitus. Dell Publishing Co. New York 1967, and J D Thompson: Organizations in Action. McGraw Hill 1967.
3. M Lonnroth, T B Johansson, P Steen: Energy in transition. Secretariat for Future Studies. Stockholm 1977.
4. PM "The Electricity Sector in Politics - History and Futures (in Swedish). Secretariat for Future Studies. To be published later.
5. Lars Bergman, Harry Flam: Energy and Economic Growth (in Swedish). Secretariat for Future Studies. Stockholm 1976.
6. Lars Lundgren: Energy Policy in Sweden 1890-1975 (in Swedish). Secretariat for Future Studies. Stockholm 1978.
7. See for example J M Blair: The Control of Oil. Pantheon Books, New York 1976. A Cockburn, J Ridgeway: Selling the Sun. The New York Review of Books, March 31, 1977.
8. Gustaf Olsson: Municipal Energy Supplies - Technology and Politics in Cooperation (in Swedish). Vasteras 1974.
9. Research, Development and Demonstration in the Energy Sector - A Global Survey 1977 (in Swedish). Energy Research and Development Commission Report No DFE-4. Ds I 1977:6. Ministry of Industry, Energy Research and Development Commission.
10. Thomas B Johansson, Peter Steen: Solar Sweden. Secretariat for Future Studies. Stockholm 1978.

11. W Greider: US Oil Industry Stakes Out Role for the Future. Washington Post. 1977-05-22.
12. The Annual Swedish Budget Bill 1970, Annex 15.
13. Power Development 1975-1990 (in Swedish). Central Operating Management. A Joint Organization of Major Swedish Power Producers (CDL). September 1972.
14. Electricity Consumption in Sweden 1975-1990 (in Swedish). CDL Study, August 1977. (CDL = Central Operating Management. A Joint Organization of Major Swedish Power Producers).
15. Svenska Dagbladet, January 24, 1978.
16. Per Hedvall: Energy Supply in Sweden towards the End of the Century - A Realistic Alternative (in Swedish). ASEA. February 28, 1978.
17. Water Forecast 1975-2000 (in Swedish). Swedish Water & Sewage Works' Association. October 1975.
18. Sven Lalander, Pricing of Electric Energy (in Swedish). Annex 6 to Policy Instruments for Future Conservation of Energy. Ds I 1977:15, Ministry of Industry, The Energy Commission.
19. G Mills: Demand electric rates: a new problem and challenge for solar heating. ASHRAE Journal, January 1977.
20. Charlotte Laren: Survey of Various Policy Instruments in the Energy Field Employed in Sweden (in Swedish). Annex 1 to Ds I 1977:16, Ministry of Industry, The Energy Commission.
21. See, for example, The Commission on Electric Heating. DsI 1977:9. Restrictions on Heating with Electric Radiators.
22. Wind Energy in Sweden (in Swedish) National Swedish Board for Energy Source Development (NE). NE 1977:2. Liber. Stockholm 1977.
23. Energy Supplies to Urban Areas and Heavy Industry (in Swedish). Swedish National Industrial Board (SIND). SIND 1976:3. Stockholm 1977.
24. Sheldon Novick: The Electric Power Industry. Environment, Volume 17, No 8, November 1975.

The Transitional period - on the energy supply in the 80's

1. Sweden's Use of Energy in the 1980's and 1990's (in Swedish). Swedish National Industrial Board (SIND). SIND 1977:9. Stockholm 1977.
2. The Energy Commission, Group C. Ds I 1978:12.
3. Energy Research and Development-78 - Energy Production (in Swedish). Prepared by the National Swedish Board for Energy Source Development. SOU 1977:61.
4. Energy Conservation (in Swedish). Main report of an Expert Group to the Energy Commission. Ds I 1977:10. Ministry of Industry, The Energy Commission.
5. Heating Plan for Stockholm - Proposal (in Swedish). The Stockholm Energy Board. Stockholm, September 15, 1976.
6. Is Night Accumulation Economically Justified? (in Swedish). Swedish State Power Board. April 5, 1977.
7. Energy, Health, Environment (in Swedish). SOU 1977:67.

Organization of Energy Policy

1. Lars Lundgren: Energy Policy in Sweden 1890-1975 (in Swedish). Secretariat for Future Studies. Stockholm 1978.
2. Carl-Johan Engstrom: Land Use and Future Energy Systems (in Swedish). National Physical Planning. Basic data No 12-77. Ministry of Housing 1977.
3. How to Save Energy. Newsweek, April 18, 1977, pp 70-80. Conservation Investments as a Gas Utility Supply Option. Preliminary Analysis. Prepared for the Federal Energy Administration of ICF Inc. January 7, 1977. FEA/G-77/010.

Choosing a Future - Uncertainties and Values

1. Fred Cottrell: Energy and Society. The relation between Energy, Social change, and Economic Development. McGraw Hill Book Co., Inc. New York, Toronto, London 1955.
2. Lynn White Jr: Technology Assessment from the Stance of a Medieval Historian. Technological Forecasting and Social Change 6, 1974, pp 359-369.
3. Technology Assessment (in Swedish). Ds Ju 1975:12. Ministry of Justice.
4. Discovery of Consequences (in Swedish). Lectures and discussions at a seminar on Technology Assessment, Norrtalje, 5-6 April 1976. Secretariat for Future Studies, Stockholm.
5. Bergstrom-Norberg-Steen: Technology Assessment. Swedish National Defence Research Institute. FOA 1 Report.
6. Harry Otway, Kerry Thomas: Understanding Public Attitudes Toward Nuclear Power. IAEA/IIASA, Vienna. Stencil November 22, 1977.
7. Olof Petersson: Election Studies. Report 2. Electors and the 1976 Election (in Swedish). Central Bureau of Statistics 1977.
8. Daniel Yankelovich: The Impact of Scarcities: The United States, Sweden, and the Federal Republic of Germany. New School for Social Research, New York, PM February 23, 1977.
9. Hans Zetterberg, Karin Busch: Priorities Assigned by the General Public to Public and to Private Consumption (in Swedish) in "Growth of Public Sector". The Swedish Industrial Council For Social and Economic Studies, 1977.
10. Ronald Inglehart: The Silent Revolution - Changing Values and Political Sytles Among Western Publics. Princeton University Press, Princeton, USA 1977.
11. Amory Lovins: Soft Energy Paths: Toward a Durable Peace. Ballinger 1977.
12. Denis Hayes: Rays of Hope. W W Norton, New York 1977.
13. Alvin A Weinberg: Social Institutions and Nuclear Energy. Science (177) 27-34, 7 July 1972.
14. Torsten Hagerstrand: To Create Coherence in the World of Man - The Problem. Man in the Technological Society (in Swedish). Lectures and discussions at the Conference of the Royal Academy of Letters, 25-27 January 1977.

The Secretariat for Futures Studies - Organization

Background

The discussion in Sweden during the later 1960's and early 1970's on the use of futures studies in governmental work lead to the setting up in 1971 of an ad hoc working group under the chairmanship of Cabinet Minister Alva Mydral. In 1972, this group published an offical report entitled "To choose a future - background material for discussions and considerations on the role of futures studies in Sweden".

On the recommendation of the working group, a Secretariat for Futures Studies was set up in February 1973. It was attached to the Prime Minister's Office and was commissioned to continue the investigations on the role of futures studies based on the responses to the first report.

Some changes were made in the organization of the Secretariat in accordance with a decision taken by Parliament in 1975. The Government now appoints the board of the Secretariat, consisting of members of parliament. In 1976, Kerstin Aner, M.P. was elected Chairman of the Board. All five parties in Parliament are represented on the Board. The Head Secretary of the Secretariat, Professor Lars Ingelstam, is also on the Board.

Projects

In 1974, Parliament for the first time allocated funds for different projects. The following four projects were initiated in 1975 and were completed in 1977-78.

- Energy and Society
- Resources and Raw Materials
- Working Life in the Future
- Sweden in the World Society

The Secretariat for Futures Studies carries out the decisions of the Board by commissioning studies.

The final report from the project "Energy and Society" is the second project report to be published. Final reports from the "Resources and Raw Materials" group and from the group on "Sweden in the World Society" were also published in 1978.

English translations of these studies are edited by Pergamon Press.

The involvement of the Secretariat for Futures Studies in energy questions began with a commission given by the government in the spring of 1974 to the then Delegation for Energy Policy and the Secretariat for Futures Studies. The commission involved the very broad mandate to conduct a futures study of the energy field.

The responsibility subsequently passed to the Secretariat. Work began in earnest in the summer of 1975. The project design was submitted as a conceptual outline which was published in the autumn of 1975. From the very beginning it was established that the project should be completed well before the decisions on energy policy which were to be taken in 1978.

A new set of projects has been started in 1978; on "Care in Society" and on "Sweden in a New International Economic Order".

Project Groups

A number of persons have been engaged for each of the projects. One person has been appointed as project leader to be responsible for the work of the group, for keeping time-schedules, planning the budget and to be generally responsible for the project. The project leader for "Energy and Society" has been Mans Lonnroth. Other members of the group have been Lars Bergman, Bo Diczfalusy, Margareta Granas, Thomas B Johansson and Peter Steen. Hans Esping has been closely related to the work. Throughout the project Erik Grafstrom has given much advice on the overall design of the project.

The independent status of the project groups and the fact that they are solely responsible for the project does not prevent close and informal contacts with the Secretariat.

An inter-departmental reference group has been attached to each of the project groups. The reference group has no responsibility for the work of the project group and only serves as a two-day channel for information between the group and the ministries concerned with the subject of the study.

Index